THE MAN WHO INVENTED
THE TWENTIETH CENTURY

Nikola Tesla 1856–1943 (at age 35).

THE MAN WHO INVENTED
THE TWENTIETH CENTURY

Nikola Tesla,
Forgotten Genius of Electricity

Robert Lomas

HEADLINE

First published in 1999
by HEADLINE BOOK PUBLISHING

10 9 8 7 6 5 4 3 2 1

British Library Cataloguing in Publication Data

Lomas, Robert
The man who invented the 20th century
1. Scientists 2. Science
I. Title
809.2

ISBN 0 7472 7588 2

Typeset by Palimpsest Book Production Limited,
Polmont, Stirlingshire
Printed and bound in Great Britain by
Clays Ltd, St Ives plc, Bungay, Suffolk

HEADLINE BOOK PUBLISHING
A division of Hodder Headline PLC
338 Euston Road
London NW1 3BH

contents

Dedicated to Elaine Ann,
for her patience, support and encouragement.

acknowledgements

I have not written this as a scholarly study because I wanted to write the story of Nikola Tesla in an easy-to-read form. He is a forgotten hero of electrical engineering and deserves to be remembered. I have not inserted footnotes or references to the people who have helped me with information and advice, or the books, magazine and newspapers I have used. I am however extremely grateful to them all.

In particular I would like to thank: Mr Tony Heyes and Drs Michael Hampshire and Robert Tomlinson of Salford University for teaching me about Ohm's Law and how to keep one hand in my pocket. They introduced me to Tesla's work and first kindled my interest in him; the late Dr John Crooks who shared a research lab with me for many years and lived with me through some exciting studies of magnetic resonance effects while still remaining my friend; Mr Gordon Brown, retired Chief Engineer of PIRA, who first put the idea of a biography of Tesla in my head and has been a prolific writer to me on the subject of Tesla's work; Mr Paul Hover who made me realise that Tesla's work deserves to be remembered and brought me up to date on its latest applications; Jenny Finder and her Library Staff at Bradford Management Centre who are always helpful, no matter how unreasonable or vague my request or how out of print the volume I want; my colleagues at Bradford University who have made many helpful comments about Tesla and his work over endless cups of coffee in the Senior Common Room; and finally my agent Bill Hamilton of A.M. Heath and my editor Doug Young of Headline Books, two gentlemen with the rare talent and patience to take the tangled skein of a first draft, find both ends and gently tease out the thread of the story from the knots of my thoughts.

Promising Beginnings?

What I had left was beautiful, artistic and fascinating in every way; what I found was machined, rough and unattractive. Is this America?

It is a century behind Europe in civilisation.

Nikola Tesla, 1884

One fine summer morning in 1884, the Calais train was about to leave Paris. Beside it on the platform stood a tall thin young man, with a head of thick black hair parted in the centre. His fashionable moustache only partly hid the nervous movements of his lips as he searched through the pockets of his suit with the urgency of somebody who has lost his wallet. He had, in fact, lost more than his wallet: all his luggage, including his small remaining savings and the ticket to New York he had only just been able to afford, had been stolen. As he realized his plight, he heard a whistle blow, the doors slam shut and he smelled the smoke and steam of the boat-train as it hissed slowly forward.

What was he to do? There was nothing left for him in Paris. He'd sold all his possessions, walked out of his engineering job and abandoned his lodgings. He closed his eyes in despair and, fortunately being blessed with a perfect photographic memory, saw the number of the steamer ticket in his mind's eye. Could he, with this information intact, still get to New York? Surely, he reasoned, the shipping company would have a record of that ticket number and he could claim his berth. As he agonized, the train started to inch slowly forward. If he missed the boat, his chance of getting to the 'Land of Golden Promise' would be lost with his belongings.

The decision made, he turned and ran along the platform, using all the length of his thin legs to catch up with and struggle aboard the accelerating train. On the long journey to the coast, he had plenty of time to check his memory for all the details of his missing ticket.

The man with this remarkable memory was a twenty-eight-year-old Serb by the name of Nikola Tesla. We know what happened that day because, in later life, he wrote a series of articles recalling the events of his early life, including the story of his lost luggage.

Simply knowing the number of the ticket, he discovered on reaching the port, was not enough to convince the steam ship company that he should be allowed on board. Taking stock of his situation once again, he searched through his pockets. There were a few coins, a handkerchief, some poems and articles he had written, a neatly tied package of calculations relating to solutions of an unsolvable integral, a rough plan for a novel type of flying machine, and a letter.

The letter was from an English friend with whom he had often played billiards in Paris. This friend, Charles Batchellor, who happened to know the famous American inventor Thomas Edison, had suggested to Tesla that America was the place to make his name as a scientist and had offered to write introducing him to Edison. Tesla had taken more care of this letter than he had of his wallet and luggage. Perhaps now he could use it to help him out of his present difficulty. Carefully opening the letter, he read:

> To Mr Thomas Edison Esq . . . The bearer of this letter
> is Mr Nikola Tesla . . .

Here was proof of his identity. He showed the letter to the embarkation officials and, when no other Nikola Tesla arrived to claim the ticket, he was finally allowed up the gangplank.

The crossing was smooth, but Tesla's discomfort can be easily imagined. His meals were provided for and he had a cabin to sleep in, but he had nothing but the clothes he stood up – not even a change of underclothes. This was a new experience for a fastidious, cultured, highly educated European gentleman. As the days at sea passed, he must have become more and more embarrassed by his unavoidable lack of personal hygiene. So aware was he of his growing personal aroma that he spent much of the voyage sitting at the stern of the boat, hoping that his body odour would be diluted by the sea air. A strong swimmer, he kept his spirits up with the hope that if a millionaire fell overboard, he could rescue him and be rewarded. It didn't happen and, for most of the way across the Atlantic, he continued to sit on deck enduring rather than enjoying the fresh air.

He was on his way to seek work with the world's most famous inventor, an all-conquering and successful entrepreneur who had taken the world by storm. Tesla had first seen the name E-D-I-S-O-N as it was spelled out, letter by glowing letter, on a motor-driven sign above the Edison pavilion at the Berlin Health Exhibition.

Hungry for recognition of his own technical skill and revolutionary ideas, Tesla wanted to tell the famous inventor, who was twenty years older than himself, that the ideas he was carrying in his mind could transform the fledgling electricity industry. His hope was that when he explained his new theory about alternating current to the great scientist, Edison would be eager to fund more research.

A well established electrician, Edison had already invented many marvellous devices that the world was clamouring to buy, including the phonograph – the first record player that played recordings made from wax cylinders – and an electric light, which he had also patented.

Edison, it seemed, was a man who was able to make money from ideas; and he'd seen a chance to get rich from electric lighting when he had realized that people didn't just want to buy a light bulb, they wanted a full electrical system, wanted to replace gas lights with the easier to use electric lights throughout their homes and businesses. At the time, when Edison set out to build an electric lighting system to replace gas, the gas lighting industry of America was worth a fortune, earning about $150 million a year.

Tesla had learnt first-hand about Edison's ambitious approach to electric lighting while he had been working for a subsidiary company in Paris, called Continental Edison. He had read an interview by Edison in a Paris newspaper that had said:

I needed to effect exact imitation of all done by gas, to replace lighting by gas by lighting by electricity. To improve the illumination to such an extent as to meet all requirements of natural, artificial and commercial conditions.

He knew that Edison had a reputation as a man who got things done; that he'd started out as a humble telegraph operator who had seen a way to improve the telegraph system; that he had made a machine, called a telegraphic relay, to receive a telegraph message and then re-transmit it on to the next station without needing a human operator. This successful invention had increased the distance a message could be sent and had saved a lot of money in operators' wages, as well as making it less likely that any operator could make a mistake copying a message by hand.

Tesla admired the way that Edison had gone on to develop a device to carry four messages over a single telegraph line, and now his only reason for travelling across the Atlantic was to meet this Wizard of the Electric Light.

Colourful stories about Edison's famous exhibitions had spread to Paris, brought by visiting American engineers. Tesla knew about the Negro attendant with the hat which lit up when he handed visitors a leaflet; and about the hundreds of men, each wearing a lamp on his head who had marched down Fifth Avenue behind a mounted marshal waving a baton with an electrically-lit tip; and about the showgirls dancing in electrically-lit costumes. Any man, he thought, who could afford such expensive displays had to have money to spare and Charles Batchellor had hinted that if he went to America, Edison might be persuaded to provide capital for him to develop the new electric motor he had invented.

As the penniless Tesla sat on the stern of the steamer, this vision of the money that Edison would give him to develop his own ideas in electricity kept his spirits up.

Disembarkation at New York must have come as a shock to Tesla. Having visited Prague, Budapest, Berlin and Paris, he had seen capital cities in Europe, but was unprepared for the raw crudeness of New York. The sweeps and curves of European architecture had fascinated him, his poetic nature responding to Gothic curves and soaring spires, but New York was a dirty crowded metropolis with buildings as crude as upturned boxes. It was rough-hewn, unfinished and teeming with people rushing about on their own business. They spoke with strange accents and drawling vowels. Standing at the end of the gangway with just four cents in his pocket, he must have wondered how he was going to survive in this coarse threatening place. He didn't even know how to find Thomas Edison of The Edison Electric Light Company to deliver his letter of introduction. He knew the offices were on Fifth Avenue, but where was that?

Setting off from the docks, he was lost within minutes. Spotting a man in uniform – obviously a New York police officer – he went up to the man and asked to be directed to the offices of Mr Edison. Struggling to make himself heard over the noise of the street, the officer shouted an aggressive incomprehensible reply. None of the dozen languages that Tesla spoke fluently could help him at this moment, and, conscious that he must look like a tramp after so long without a change of clothes, he nervously repeated his question. Waving his big stick, the cop pointed down the road.

Thanking him for his help, Tesla set off. New York, with all its bustling activity, was very different from the more sedate European cities he knew. In European cities he felt at home, knew their heritages and architectural form, here he was an alien. Even the language was different from the formal English he had learnt at school. He was disorientated, dirty and hungry, with only four cents in his pocket. It was late morning by now, and he was beginning to wonder how he would pay for his next meal.

In 1917, in a speech he gave to the American Institute

of Electrical Engineers, when he was awarded the Edison Gold Medal, he recalled those first few hours in New York, telling how, as he walked up the road, he passed the open door of a small workshop and, for the first time in America, saw a familiar object. There, in the workshop, was a large dynamo of a type that he had installed in both Paris and Stuttgart. A man was working on it and, from his tone of voice, it was clear he was having trouble.

Unable to resist stepping inside for a closer look at the machine, Tesla asked what the trouble was. The man told him that foreign dynamos were impossible to repair; that it was a European machine and he couldn't find out anything about it. Taking off his jacket, Tesla, glad of a chance to do something useful, offered to help. By late afternoon he had repaired the machine, and the man had promptly offered him a job as a repairman. Tesla had politely refused, explaining that he was on his way to a job with Mr Edison. The man thanked him for his help and gave him $20. Surprised and pleased, Tesla was now able to afford a bed for the night and, before leaving, he made sure that he had detailed directions for getting to Fifth Avenue the next morning.

'You can't miss it,' the repair man had told him. 'Nobody but Edison would put so many striped window blinds on a mansion.'

The following day – refreshed, restored and wearing newly bought clean underwear and socks – Tesla went to Edison's Fifth Avenue headquarters, number 65 at the western end of a terrace of large fashionable houses. The repairman had been right, it stood out from the other elegant town-houses because it was the only one with a gaudy striped sunblind above each of its three rows of four south-facing windows.

As Tesla walked up the avenue he could hear the rumbling of a steam-engine coming from the small engine house that had been built on to the western end of the basement. Stopping to admire the pillared porchway, he must have wondered why that elegant porticoed doorway also needed a matching gaudy striped shade to keep the sun off the door. The ground floor was well above street, level with a wide sweep of fifteen steps leading up to it. It was an impressive office, its bright striped shades catching the eye during the daytime and its electric lighting shining out during darkness. Tesla stood for a moment on the top of the steps and looked along Fifth Avenue. He was standing on the same spot where Edison had stood to review a great parade of electrically-lit men, and had no way of knowing that soon people would be acknowledging him with similar honours. He went inside, produced Batchellor's letter and asked to see Mr Edison.

Ushered into Edison's office, Tesla saw a cluttered room, lined with shelves, containing a workbench as well as a desk, and brightly lit by electric light. The next surprise was how homespun Edison looked in his dark three-piece suit, white shirt and black bow tie. He certainly did not look the larger-than-life hero that Tesla had expected; seeming more like a well-to-do farmer dressed up for Sunday Church Service. A few inches shorter than Tesla, he had thinning grey hair and rounded shoulders, but his clean-shaven face had a look of assurance about it that suggested he was used to getting his own way. Greeting Tesla warmly, he read Batchellor's letter and, without any further ado, offered him a job.

Accepting on the spot, Tesla started to tell Edison about his mathematical predictions of how Alternating Current could be made to work an entirely new type of motor. Edison listened for a while, but it was clear he was preoccupied. Eventually he told Tesla that he was not interested in any new theory of electricity. He had a system that did its job, which he had worked out for himself, without any need to fuss about with mathematics.

Tesla was impressed by this powerful man who, without any social advantages or scientific training, had accomplished so much. With uncharacteristic self-doubt, he wondered if he had been wasting his time studying mathematics, science, literature, art and a dozen languages. Perhaps he needn't have spent so long in libraries reading everything from Newton's *Principia* to the novels of Paul de Kock. If Edison could become so successful without any theoretical training at all, what use was academic study?

For a moment, Edison's obvious lack of interest in the theory of electricity made Tesla doubt all his book learning and education, and he knew that he would have to win the man's confidence by proving himself to be a practical engineer. Only then would Edison be prepared to listen to his ideas for the future of electricity. Then, as Edison talked, Tesla saw a way to demonstrate his engineering skill.

Edison had been commissioned by the famous naval architect, Louis Nixon, to fit a lighting system to the *SS Oregon*, a passenger ship built for the transatlantic run. Two generators had subsequently been fitted, but were too large to remove from the ship. Unfortunately both the main and the reserve dynamos had failed at the same time and consequently the ship could not sail.

Edison was also preoccupied at that time with news he had just received that his wife had been taken ill with typhoid. Understandably, he was in no mood to listen to the radical theories of a young Serbian engineer.

Tesla reminded Edison that he had already proved himself capable of sorting out installations while working for Continental Edison in Paris. He asked for the chance to repair the *Oregon*, and was delighted and surprised when Edison agreed.

Before electric lighting units were invented, oil lamps and candles were the only way to light ships at sea. Gas lamps couldn't be used because there was no way of making or storing enough gas on board. An electric lighting system was much more convenient and effective than oil lamps, so when Edison started to sell free-standing lighting generators, ship builders saw an opportunity to improve their ships. Louis Nixon had built the *SS Oregon*, the fastest, most comfortable modern liner on the Atlantic crossing, and had made use of all the latest scientific advances, including an Edison lighting system.

On its first voyages, the new lighting system was hailed as a triumph and a breakthrough in marine lighting, but in the late summer of 1884 the lights on the *Oregon* had gone out and would not come back on again.

When Tesla went on board the *Oregon* he found that both dynamos had burnt out wiring in their main coils. Edison's staff said they could only be repaired at the factory, but the dynamos were too big to be removed from the ship's hold. Installed as the ship was being built, they would not come though the finished hatches. Although somewhat disturbed that nobody had bothered to calculate the size of the hatches to allow for easy servicing of the generators, Tesla had no time to worry about that now. The ship could not sail without lights and the owners were losing considerable sums of money while the ship was idle. Wasting no time, he organized a working party of seamen, stripped the machines down and, working through the night, rebuilt them. By five o'clock the following morning, the ship was seaworthy.

Feeling he had proved himself to Edison, Tesla was tired but pleased as he walked towards the city from the docks. It was 5.30 a.m., and the streets were quiet when he bumped into Edison. To his surprise, Charles Batchellor was with him. By way of greeting Tesla, Edison said sarcastically to Batchellor, 'Here is our Parisian running around at night.'

He was, however, pleasantly surprised when Tesla explained that he had repaired both of the *SS Oregon*'s generators. And Batchellor's initiative in recommending Tesla certainly met with Edison's approval, for as the two men walked away Tesla heard Edison comment, 'Batchellor, you've got a dammed good man.'

chapter I

Destined for the Priesthood?

The gift of mental power comes from God, Divine Being, and if we concentrate our minds on that truth, we become in tune with this great power.

My Mother had taught me to seek all truth in the Bible.

Nikola Tesla

Revd Milutin Tesla's son Nikola was born, in the small village of Smiljan in Croatia, during a spectacular thunder storm on the stroke of midnight as July 10 turned into July 11, 1856. The midwife, who attended his mother Djouka, was so frightened by the lightning that she said the baby must be a child of the storm. She could not have known how accurate this description would be for a man who was destined to create artificial lightning strong enough to shake the whole world.

During his long lifetime – he died in 1943 – Tesla would relate many anecdotes about his childhood, both in his speeches and in various writings. From such records, we know something of his early thoughts and development.

Tesla always claimed that his interest in electricity began one cold winter's day when he was three years old and experienced the thrill of drawing static electric sparks from the fur of Macak, his pet cat. This strange effect, caused by the mixture of intense cold and dry air, started him wondering about natural electricity: the lightning of storms. 'Is nature a gigantic cat?' he thought. 'If so, who strokes its back? It can only be God.'

This childish curiosity stayed with him throughout his long career. Recalling the incident in his eighties, he admitted he still asked himself daily, 'What is electricity? Eighty years have gone by and I still ask the same question, unable to answer it.'

Nikola Tesla, the fourth child and second son of Revd Tesla, was known affectionately to the family as Niko. The Teslas were a close family and Niko adored his big brother Dane, who was seven years older than him. Dane was a cheery lad, clever at literary subjects, and his parents were looking forward to him doing well at school and then following his father into the priesthood. It was not to be.

Revd Tesla had a fine Arab horse that he was extremely fond of. The animal had once saved his life during a snow storm by leading rescuers to him after he had fallen off. This horse became a great pet of the family and all the children rode it. One day, while twelve-year-old Dane was riding the horse, it slipped. Dane fell off and was trampled beneath its hooves. Nikola, who was five at the time witnessed his brother fall and then die soon afterwards. This tragedy made a lasting impression on him and, years later, he said of the incident:

> *This horse was responsible for my brother's injuries from which he died. I witnessed the tragic scene and, although so many years have elapsed since, my visual impression of it has lost none of its force. The recollection of his attainments made every effort of mine seem dull in comparison. Anything I did that was creditable merely caused my parents to feel their loss more keenly. So I grew up with little confidence in myself.*

Revd and Mrs Tesla were heart-broken. Niko was now their only surviving boy, having two older sisters and one younger one (Marica, Angelina and Milka). The pressure to replace his brilliant, much loved brother, drove Niko to work exceedingly hard to achieve great things. A career in the priesthood, however, did not interest him. He wanted to know how the world worked and to change it for the better. He could not compete with his dead brother in literary subjects, but he was a natural mathematician. Indeed, he was so quick at mental arithmetic that, when he started school, his teacher thought he must have seen the answers she had written down. The only way he could think of making his parents proud of him was to use his ability in maths to become a famous scientist and inventor.

Niko was always inventive. His first successful invention was a frog-catching hook that he made when he was about six. One of his playmates was given a fishing rod and decided to test it out catching some of the frogs which swarmed around the village. Having quarrelled with the fishing-rod owner, Niko was excluded by the village children from the frog-fishing expedition. He decided to make his own equipment and go on a lone fishing trip. Finding a piece of soft wire, he formed a sharp point by hammering the end between two stones, bent it into a hook shape, and attached it to a length of string. But none of the frogs in the stream would bite, even though he tried all sorts of attractive baits.

Feeling pretty fed up, he noticed a frog sitting on a tree stump by the bank of the stream, and swung the empty hook in the air in front of it, just to see what would happen. The frog was irritated by the swinging hook, lunged at it and stuck fast. Young Niko had found a simple method of catching frogs. Meanwhile, his playmates, who had had no success with their fine rod and tackle, were green with envy. Niko teased them for some time, and wouldn't tell them how he had done it. He just paraded his large catches of frogs to annoy them. Eventually when the quarrel was made up, he shared his secret, much to the detriment of the local frog population.

Even as a very young boy, mechanical things interested

him. His family moved to the little town of Gospic where they were invited to the first display of the town's new fire engine and, along with the rest of the townspeople, Niko felt let down when the machine failed to pump any water from its hose. The strenuous pumping efforts of a vigorous team of local volunteers couldn't make the hose squirt water. The adults stood about bewailing the fact, but young Niko dived into the river and opened up the collapsed suction pipe, letting the water run to the pump. The sudden flowing jet came as such a surprise it showered many of the assembled local worthies before it was brought under control.

Nikola Tesla went to school at the Real Gymnasium in the town of Carlstadt, Croatia, before going on to read engineering at Graz Polytechnic and Prague University. He was an exemplary student, remembered as almost working himself to death. He never seemed to do anything other than work and, for amusement, would set himself complex mathematical problems to work out, saying those sort of sums relaxed him. He was keenly disappointed when his father made light of his early academic successes.

Soon after his graduation his father died and Nikola had to sort out his father's estate. While sorting through his papers, he found a series of letters from his own professors who had become so concerned about him that they had written to Revd Tesla asking him to dissuade his son from working too hard. Perhaps if his father had been less afraid of praising him, Nikola would not have been so easily influenced in later life by substitute father-figures who exploited his vulnerability and eagerness to please.

While still young, Nikola devised some awesome machines in his mind and enjoyed working out all their technical details. He suggested an under-water tube to carry letters and parcels under the seas in globes strong enough not to collapse under the water pressure. He worked out the necessary power for the pump to force the water through the tube. This would be an enormously fast method of sending messages, he decided. But then, after a lesson about water friction in pipes, he realized his idea needed far too much energy to work.

While at Carlstadt, Nikola, a precocious and independent youngster, extended his interest in electricity. Once he had set his mind on an idea he could not be diverted. He was determined to be an engineer and believed he would be an outstanding one. Once when recovering from a childhood illness he had told his father that he did not think he was capable of becoming a priest, but that he loved maths and engineering. His father, worried by Nikola's tendency to overwork and make himself ill, nevertheless told him he could be a scientist if that was what he wanted.

Nikola was a keen and adventurous swimmer. On at least two occasions he nearly drowned while swimming in fast-flowing currents. Learning from these adventures he became interested in the power of flowing water. How, he wondered, could he turn that flow into a turning force? He calculated how much power he could hope to generate from the energy of the flowing river and was amazed. From then on, he saw the power of flowing water as a free source of energy if mankind only had the wit to use it.

Although highly educated academically, Nikola had no head for business, perhaps because his family had never been in business. His father, the son of an officer in Napoleon's army, had been given a military education before joining the clergy. One of his uncles was a professor of mathematics and another a colonel in the army. Nikola was brought up to respect knowledge for its own sake, not for what it could earn.

When he went to the prestigious Real Gymnasium at Carlstadt it was too far from his father's parish in Gospic for him to commute, so he had to live away from home. As the school didn't take boarders, young Nikola moved in with his uncle, the retired army colonel.

Carlstadt, a low marshy part of Croatia, buzzed with mosquitoes and was subject to malaria. Soon after his arrival, much to the concern of his aunt, Tesla caught the disease. Believing that further bouts of illness could be avoided by ensuring that his digestive system was never overloaded with food, she kept young Nikola in a state of perpetual hunger. Later, he said of that time:

My aunt was a distinguished lady, the wife of a colonel who was an old war-horse having participated in many battles. I can never forget the three years I passed at their home. No fortress in time of war was under a more rigid discipline. I was fed like a canary bird. All the meals were of the highest quality and deliciously prepared, but short in quantity by a thousand per cent. The slices of ham cut by my aunt were like tissue paper. When the Colonel would put something substantial on my plate she would snatch it away and say excitedly to him: 'Be careful. Niko is very delicate.' I had a voracious appetite and suffered like Tantalus. But I lived in an atmosphere of refinement and artistic taste quite unusual for those times.

These privations, plus the military rules in his uncle's house, helped Tesla to develop self-control and consolidated an unhealthy respect for authority figures which stayed with him for the rest of his life. His characteristic distrust of women was reinforced by the starvation tactics of his formidable aunt, but he learnt by example how the application of will-power could achieve goals.

Most people have one inspiring teacher in their past and Tesla was no exception. At Carlstadt he had the good luck to be taught science by an extremely gifted enthusiast, Professor Poeschl, who was forever seeking out new experiments to inspire his young charges. He taught theoretical and experimental physics. Tesla described Poeschl in his memoirs as:

. . . an ingenious man who often demonstrated the principles by apparatus of his own invention. Among these I recall a device in the shape of a freely rotatable bulb, with tinfoil coating, which was made to spin rapidly when connected to a static machine. It is

> impossible for me to convey an adequate idea of
> the intensity of feeling I experienced in witnessing
> his exhibitions of these mysterious phenomena. Every
> impression produced a thousand echoes in my mind. I
> wanted to know more of this wonderful force.

One lesson really impressed Tesla. Poeschl brought in one of the latest electrical dynamos from Paris to show his pupils how a generator could also work as a motor.

The earliest sources of electricity were batteries which only make direct current; so, naturally, when the first electric motors were made, they used batteries to power them. When Michael Faraday made the first electric generator, he found that if he allowed a coil of wire to spin round in the field of a magnet then a strange alternating current would flow through it. But nobody could make the strange reversing current drive a motor. Then it was found that if mechanical switches were fitted on the end of the coil, the current could be turned backwards and forwards so that it worked like a direct current. The Gramme dynamo, which Professor Poeschl bought in Paris, used these switches in a device called a commutator.

The machine could be used either as a dynamo or a motor, but when Poeschl ran it as a motor Tesla noticed that the metal brushes that connected the electricity to the moving commutator kept sparking and crackling. He said what he thought: 'Surely, Professor, it should be possible to run a motor without using this method.

See how badly the brushes spark. It would be much more efficient without them.'

All good teachers treat their pupils with respect and Professor Poeschl was no exception. He took Tesla's comments seriously and devoted the next lesson to explaining why alternating current could not work an electric motor. At the end of his lecture he said: 'Mr Tesla will accomplish great things, but he will never make a motor run on alternating current. It would be equivalent to converting a steadily pulling force, like that of gravity, into a rotary effect. It is an impossible idea.'

Tesla, however, was sure it could be done. He reasoned that if an alternating current came from a circular motion, then it should be possible to make a circular motion out of an alternating current.

Even the best education must come to an end. Tesla was enjoying his studies, but after the death of his father realized that he could not expect his mother to support him for ever. He needed a job.

Mr Puskas, a good friend of his late father, was running a company installing a telephone system in Budapest, and was persuaded to take Tesla on, so he moved to Budapest. Still convinced he could make an alternating current motor, Tesla drove himself so hard with the craze to make a new type of motor that he had a nervous breakdown. His main symptom during this time was an extraordinary acuteness of hearing that made all noises extremely distressing for him. Throughout his life, Tesla's hearing was always acute, but, during his breakdown, it became so sensitive that he believed even the sound of fly landing on a table made a painfully heavy thud.

Placing his bed on rubber cushions to give himself some relief from the vibrations of the city, he continued to claim that he could hear all the surrounding conversations of the town in a great roaring jumble of sound. Many years later he said that the only thing that kept him going during this period was his desire to make a rotating magnetic field using alternating current. Of his passion to create this new type of electric motor, he said: 'With me, it was a sacred vow, a question of life and death. I knew that I would perish if I failed.'

Perhaps this obsession caused his breakdown. Certainly – although it could have been pure chance – the solution to his alternating current problem coincided with the cure of his illness. During the worst periods of sickness, he said the answer had hovered at the back of his mind, just out of reach.

Writing about this many years later in a collection of articles, published posthumously as his 'autobiography', he described the revelation which came to him as he was walking through Budapest City Park with a friend. He was watching the sun set and reciting aloud from Goethe's *Faust*:

> *Like a flash of lightning and in an instant the truth was revealed. I drew with a stick on the sand the diagrams of*

> *my motor. A thousand secrets of nature which I might*
> *have stumbled upon accidentally I would have given*
> *for that one which I had wrested from her against all*
> *odds and at the peril of my existence.*

Tesla's artistic nature is revealed very clearly in the way that he tells the story of his great discovery. It came to him as a complete and breathtaking poetic thought as he gazed on the beauty of nature and enjoyed reciting Goethe. He saw himself as a great artist who works with science and nature, but who is above the mean ways of the world.

His idea was beautiful. Nobody before had made an alternating current (AC) motor. When other engineers had tried, they found that the magnetic fields produced by alternating current just churned about, not turning the motor. The magnetic field died when the current reversed direction and so the motor stopped. What Tesla did, was to use two alternating currents that were out of step with each other. Like the propelling waves of legs that move a millipede forward, the magnetic fields worked together to push the rotating shaft of the motor around. By using more than one set of currents, he could ensure that there was always a strong current available to power the motor.

As one of the currents died away, the other would continue to move the motor round. The magnetic field rotated and carried the motor round with it, and it did so without using any electrical connections to the rotating shaft. He had done what he had told Professor

Poeschl he would do – he had got rid of the inefficient sparking commutator. The electricity was connected to the moving armature of the motor by a wireless method of magnetic induction. As the current flowed through the coils, which made up the stationary part of the motor, it created a moving magnetic field which, cutting through the wires of the rotor coils, made a current flow without needing any wires to connect to the moving parts. At that moment Nikola realized that it was possible to make electricity flow through space without using wires.

The idea was simple and commercial. But Tesla, unlike Edison, never understood how to measure his work by the dollar. He was driven by the need to solve a problem and, once it was solved, he never thought about how to persuade people to pay him for his work. This was his main weakness and it was present from the very start of his career.

Over the next few months, Tesla worked out the detail for a complete alternating current system. He designed generators, motors and transformers – everything that was needed to revolutionize the long-distance transmission of electric power. His good memory and ability to visualize his designs meant he didn't produce many drawings, but, instead, stored his designs in his memory. In this extremely productive period, he devised the prototypes of many devices that would eventually be so successful in today's world. He extended his idea of using two sets of current to using three currents. This, he thought, would improve the motor's efficiency. He called his new invention the polyphase motor. All this, before he had even built his first two-phase machines. The whole system of AC machines existed only in his mind, but his knowledge of the theory of electricity made him quite sure they would work.

Tesla was a pioneer of modern electrical engineering which relies greatly on a mathematical understanding of what is happening. He studied mathematics and the work of earlier scientists so he would understand how electricity worked. He did not simply use trial and error as Edison always did. He thought about his problems and worked out how to solve them before he built any equipment. As he got to know Edison, he was not impressed with the Great Man's unscientific methods, saying of him: 'If Edison had a needle to find in a haystack, he would proceed at once with the diligence of the bee to examine straw after straw until he found the object of his search.'

Edison's skill, though, was not really electrical, it was making money from other people's ideas. Tesla was a different sort of inventor. His mind was totally original, seeing both patterns and how to use them. He was driven by an internal agenda to prove himself 'worthy', and was not prepared to accept other people's views on the limits of what was possible. If maths and logic persuaded him a thing could be done, he would be totally stubborn about proving it and would never doubt he was right. Eventually, he developed a theory of electricity which would change the world.

chapter 2

A Little Theory Goes A Long Way

*The current through a circuit is proportional to the applied
e.m.f. and inversely proportional to the resistance.*
Georg Simon Ohm, 1827

*There is scarcely a subject that cannot be mathemati-
cally treated and the effect calculated beforehand or
the results determined beforehand from the available
theoretical and practical data.*
Nikola Tesla, 1919

Tesla's mathematical view of electricity was far in advance
of most scientists in the late nineteenth century. Indeed,
his approach and thinking were so ahead of his time that
today's ways of analyzing electricity, taught in modern
university departments of Electrical Engineering, would
not have seemed strange to him.

The two most useful things a young electrical engineer ever learns are a simple formula – V=IR (Voltage = Current x Resistance) – known as Ohm's Law and how to not to suffer lethal consequences from electric shocks.

At one time, all trainee electrical engineers were taught to keep one hand in a pocket when working near live electrical equipment. The reason is simple: a shock across the chest, from hand to hand, will kill by stopping the heart. The same shock down one side of the body will only result in a severe jolt. Hence the advice to keep one hand well out of the way, in order to avoid shocks across the chest. If your hand is in your pocket, you cannot accidentally touch a live terminal with it.

The value of Ohm's law was never really grasped during Ohm's lifetime. He spent years doing tests to discover how electric circuits worked, then as long again checking the results over and over to prove his law. The scientists of his day never guessed what a marvellous tool he had given them and he got no official recognition, spending most of his life in poorly paid jobs. It was only two years before his death that he was finally made a Professor of Physics at the University of Munich. Nikola Tesla was the first engineer to successfully utilize Ohm's law to take electricity to the people.

Ohm's law foretells the future. For an electrical engineer, it gives marvellous insights, making accurate predictions about how any proposed circuit will perform, before it is even built. The current passing though a length of wire, and the voltage pressure pushing that current along, continually change. If a circuit is to work as it is intended to, a designer needs to understand how and why these changes occur. In the early days of electricity, most engineers did not know why voltages and currents changed. Ohm's law explains how electricity behaves when it flows as a current, but most early scientists studied static electricity. They weren't interested in current flow, so saw little use for Ohm's law.

Static electricity began as an innocent magic such as children make when they rub a balloon on a woollen sleeve to make it stick to a wall. This conjuring trick is an effect of 'static electricity' and was known in the time of Queen Elizabeth I of England. It was noticed by William Gilbert, the Queen's court physician and a man with many outside interests. When he was not attending to the Queen's health, he spent his spare time studying the hidden mysteries of nature and science and observed something very odd about the way little pieces of paper and other lightweight objects behave.

Gilbert found that if he rubbed a piece of amber with fur, the amber would attract small objects to it. Most school children will have repeated this famous experiment which first showed static electricity. They tear up some tiny pieces of paper and lay them on the table. Next, they take an ordinary plastic ball-point pen, rub it on the sleeve of their jackets and hold the pen over the bits of paper. The pen attracts the paper to it.

Dr Gilbert didn't have a ball-point pen, which is why he used a piece of amber when he first did the experiment. There was no name for this strange force that attracted light objects, so Dr Gilbert named it 'Electric', making up the name from the Greek word for amber, *Elektron*. Dr Gilbert studied at St John's College, Cambridge, before he became personal physician to Queen Elizabeth, and is known as the first electrician because he made up the name for it.

He studied both electricity and magnetism, discovering the earth to be a giant magnet and explaining why compass needles always point to the North Pole. He published his findings in the first scientific book written in England, entitled *Of Magnets, Magnetic Bodies and the Great Magnet of the Earth*. To honour his discoveries, one of the units that today's engineers use to measure the strength of magnets is called the Gilbert.

Even though Gilbert had written a book about electricity, for years nobody knew how to use this strange force. What the scientists of the time did not know, was that static electricity is caused by small charged particles, called electrons. These can either be added to or taken away from materials called insulators. (An insulator is something that stops electricity passing though it. Typical examples of insulators are glass, cloth or plastic.) These displaced electrons are called the 'charge'. Electrons moving about in conductors cause currents to flow. (A conductor is something that lets electricity pass through it. Typical conductors are copper, gold and silver.) A moving charge creates a current.

When a plastic pen is rubbed against a woollen jumper sleeve, electrons are pulled from the pen into the wool. This means that the pen loses some electrons and the wool is left with too many. The electrons then try to get back into the gaps left in the pen. If a small piece of paper is held near to the pen, the electrons in the paper try to fill the spaces left by the electrons lost from the pen. As the electrons try to move to the pen, they carry the paper with them.

As more and more scientists studied this strange attraction, a Frenchman, Charles Dufay, found that static electricity came in two different types. He realized that if he rubbed two pieces of amber together, and then put them close to one another they pushed each other away. So, he had discovered that if two amber insulators are both short of electrons they push against each other. In the same way, if two insulators have too many electrons they also push each other away. Only when an object with too many electrons is put near an object with too few electrons do they pull towards each other.

Benjamin Franklin, the American scientist and politician, named these two types of electricity positive and negative, and wrote down the rule that similar types of electricity push apart, but different types pull together. Nowadays, we know that objects with extra electrons are negative while those with too few electrons are positive. Rubbing an insulator to remove electrons is called 'charging' it. Once an object has been charged (by adding or taking away electrons) it can keep that charge indefinitely. Because the electric charge just sits on an object without moving, this sort of electricity is called static electricity.

Electrons that have been crowded together try to

move from where there is no room to where there is a shortage of electrons. They can only do this if there is a conductor to flow along. The moving electrons cause currents and Ohm's law explains how those currents behave.

Many people experience electrons moving suddenly. Anyone wearing nylon clothing, for example, and rubbing against carpets or seats, will get an electric shock when they next touch metal. The rubbing of the nylon clothing takes electrons away from them. When they touch any conductor, the electrons from that conductor all rush towards them at once causing the electric shock. When this happens, people blame the shock on static.

Until 1746 there was no way to store a static electric charge, then two scientists, Ewald Georg von Kliest and Pieter van Musschenbroek, made a jar for keeping electric charge in. Because they worked at the University of Leyden, their invention became known as the Leyden jar. The Leyden jar is rather like a pickle jar for keeping onions in, but the Leyden jar stores electrons in metal instead of onions in vinegar. Those two scientists made it possible to keep static electricity for later use. This was very useful in the days before batteries were invented. Today's engineers call tiny Leyden jars 'capacitors' and still use them to store charge in all sorts of electrical circuits.

The Leyden jar and the capacitor work by having two separate metal plates that can either store extra electrons or maintain a shortage of them. If you imagine a glass pickle jar covered inside and outside with metal foil, with the glass separating the two metal coatings, you will have a good idea what the jar looked like. When there is an imbalance between the number of electrons in the two plates, the jar becomes charged with electricity. If you connect the two plates together with a conductor, current flows until all the spare electrons have moved back to where they came from. When both plates have the same density of electrons, or level of charge, the current stops flowing. The greater the number of electrons that are separated and shared between the plates, the more electricity is stored. The more electricity stored, the longer the spark that can be made when the electrons are allowed to flow together.

You can understand this by imagining a dam. The build-up of electrons is like the piling up of water behind the dam. If the dam sluice is opened, the water rushes out. The higher the depth of water behind the dam, the faster and more vigorous the outflow. In the same way, the spark is longer and more vigorous when lots of electrons have been separated between the two plates.

When we talk about the strength of a flow of water, we are talking of the 'head' of water. The greater the head, the more pressure there is behind the flow. This idea was applied to electric charge and scientists started to talk about the pressure of the charge. They called this pressure the electromotive force (e.m.f.).

Scientists who make discoveries about electricity often have units of measurement named after them. Two scientists you may have heard of are André Marie Ampère and Count Alessandro Volta. Ampère, a French mathematician and physicist, invented an instrument for measuring the flow of electric charge and had the unit of current (the Amp) named after him. Count Volta, an Italian physicist, invented the first battery, the voltaic pile, and in honour of this achievement the unit for measuring electromotive force was named the Volt. Tesla, whose motor used rotating magnetic fields, had the unit of magnetic force named after him.

Static is the most spectacular form of electricity. It produces dramatic sparks, makes dead frogs' legs twitch theatrically, and causes thunder storms; but it has never really had a practical use. Useful electricity comes from Count Volta's batteries.

On 20 March, 1800, Volta wrote to the President of the Royal Society in London, saying: 'I have the pleasure of sending you some striking results at which I have arrived in pursuing my experiments on the electricity produced by the simple mutual contact of different metals. The chief result of which is an apparatus whose charge is restored automatically after each discharge.'

That was how Volta described the very first electric battery. But batteries run flat and flat batteries have no volts. Until a more steady source of electric current could be found, electric motors and lights were too expensive to use because they quickly ran down their batteries. What was needed was a continuous generator of electricity.

Michael Faraday invented the first electric generator in 1831. The electricity that Edison made from his steam generators was the same type of continuous electricity that came from Volta's batteries. Edison had quickly realized that he needed to make his lamps use less current so that he could use thinner copper wire in his distribution system. He knew that voltage drop would restrict how far he could send his electricity unless he cut down the current he used. He was, in fact, applying Ohm's law without really understanding it. It took Tesla to solve the problem of transmitting electrical power over long distances, and he did this by understanding and using Ohm's Law.

Edison had started out as a telegraph operator and this had given him an inflated idea of how far electric current would travel. His company was financed by men who had got rich from the telegraph and they expected electric current to travel at least as far as a telegraph message. It didn't. Nobody involved with the telegraph understood Ohm's Law, so they did not foresee the practical problems of using direct electricity at high currents.

Looking at the history of the telegraph, it becomes clear how they were all misled.

The same year, 1831, that Faraday invented the electrical generator, an American, called Samuel Findlay Breeze Morse, visited Britain. Morse was not really an engineer, but a well-known painter and sculptor who was awarded a Gold Medal by the Society of Arts, in London, in 1812. He was on the liner *SS Sully*, sailing back across the Atlantic after receiving his medal, when he heard from a fellow passenger about Faraday's new developments in electricity. He also heard that railway companies in England needed a better form of message telegraph and this gave him his idea for sending messages by electricity.

Morse was Professor of Fine Arts at New York University. When he got back to America, he spent all his spare time working on a system for sending messages along a wire. He did this by switching between long and short pulses of electricity. At the other end of the wire the electricity would either make a needle twitch or sound a buzzer. Morse developed a simple code, made up of short pulses (called dots) and long pulses (called dashes) to send messages. This was an elegant invention because the message seemed to arrive at the other end of the wire as soon as it was sent. It had one major drawback though: if the wires connecting the sending and receiving stations were more than 20 miles (32 km) long, then the signal became too weak to hear. To overcome this problem, telegraph stations had to be set up at 20-mile (32-km) intervals. When a message arrived, the operator would copy it down and then send it on to the next operator. The US Congress gave Morse a grant of $30,000 to set up a public message system, and on 24 May 1844, Morse sent his first public telegram from Washington DC to Baltimore in Maryland. The message he sent said, 'What hath God wrought!'

If Edison had been able to transmit electric power for similar distances, his system would have needed far fewer power stations and been more successful. But he didn't realize that the current he wanted to use was too high to travel this far.

A wire does not really like electricity going through

it, and tries to stop the electrons passing along it. First it gets hot and if you keep increasing the current, the wire finally melts. You can't force it to carry any more current than it wants to. Thick wire is more tolerant and will carry more current than thin wire, but thick wires use a lot more copper and cost more to buy. Every wire has an absolute upper limit to the load of current it will carry.

Ohm said that 'the current through a circuit is inversely proportional to the resistance'. This means that the more current you pass though a wire, the more the voltage pressure drops. If you increase the current you can force a higher voltage at the end of the wire, but the wire will melt and stop working. It's a real Catch 22. With a low current, there is less voltage drop over a particular wire than with a high current, but you need a high current to make things work. The amount of light you can get from a bulb, or the amount of work you can get from an electric motor, depends on the amount of current you can get to it.

To make a light bulb glow brightly, needs power. Power is really the ability to do work. Electricity is measured by two things, voltage and current. To make electric power needs both voltage and current to appear at the same time. The electric light will only glow brightly when the voltage and the current work together.

The more power you have, the brighter your light will shine, but there are different ways of getting high power to an electric light. You can have a high voltage and a low current, or a high current and a low voltage. The power of an electricity supply, or its ability to do the work, is found by multiplying the voltage by the current. Voltage is the ability of electricity to travel along wires. The higher the voltage, the further the electricity can go. Current is the number of individual particles of electricity (these are very small things called electrons) which actually do the work at the end of the wire.

Because power needs both voltage and current, this means that the electricity must have the ability to pass through a wire and have sufficient electrons to do the work when it gets to the lamp. The longer the wire, the less voltage there is left over at the end to do any work, but this is not the full story. If it were, then just using high voltage would make the length of the wire unimportant. The higher the voltage, the easier it is for the current to pass through the wire, but it is also easier for that current to pass through a human being. If a high voltage is accidentally connected to a person, then it will pass enough current through them to kill. This puts an upper limit on the voltage that can be used in a house because of the risk to life. The higher the voltage, the more difficult it is to protect people from it (i.e., the more difficult it is to insulate the wires).

As electrons travel along a wire they bump into the atoms of the wire and make them hot. This slows down the electrons and reduces the voltage. So the longer the wire connecting your home to the power station, the less voltage you get and the dimmer your lights are.

To take advantage of Ohm's law, electricity should be generated at a low voltage and high current (to avoid sparking and insulator break-down in the mechanical generator), transmitted at very high voltage and low current (so there is very little heating loss in the wires and less copper can be used) and then used at a low voltage in the home (to avoid the danger of electrocution). When electricity was first discovered, it was not possible to achieve all these things.

There are two types of electricity: Direct Current (DC) and Alternating Current (AC). DC is a continuous type of electricity that never changes its direction; AC is a switch-back type of electricity that is always changing and reversing its flow. The AC electricity that comes out of the socket in your house reverses its direction at least 100 times every second. AC can have its voltage changed up or down simply by passing it through a pair of coils, called a transformer. DC can only have its voltage reduced, but not increased again. The very first electric motors could only use DC because, in the mid-nineteenth century, most engineers thought like Tesla's teacher, that it was impossible to use AC for any practical purposes. As a result it was largely ignored.

At the voltages that Edison could safely use, when he started to produce domestic electricity, if you lived more than half a mile from one of his power stations your lamps would be too dim to cast sufficient light. This was a fundamental weakness of DC electricity for public electricity supplies.

To understand this, think again about that dam full of water and, this time, imagine a sluice gate. The voltage is like the head of water, and the current is the flow of water through the sluice gate. If the dam has no water in it, opening the sluice fully will not give water flow. If the dam is full to the top, opening the sluice gate just part of the way will give a strong flow.

If we want to use water to feed a hose and wash our car, then we need a flow of water with a good head to give the jet cleaning power. If the flow is too little, or the head too weak, the hose won't have enough power to wash off the dirt. If the hose gets kinked, the flow of water slows down or stops. If the head of water drops because of a leak in the pipe, the hose will not have sufficient power to clean the car. The actual power comes from both the flow and the head acting together. Electrical power works in the same way. With a high voltage and plenty of current, you get lots of power. If the voltage is lost in the transmission wire, then there is not enough pressure at the end of the wire to allow the current to do any work. By increasing the voltage, you can overcome the losses in the wire and still have enough pressure to do work at the end of it. If you can transmit your electric power at a very high voltage and use a very small current, then you get the same amount of power and much less of a problem with pressure loss in the wire. But you must be able to reduce the voltage before you let it into people's house to avoid killing them. Nowadays the National Grid wires that carry electricity between massive pylons all over the country work at millions of volts. Electricity is distributed locally at about 11,000 volts and reduced to 240 volts before it comes into our homes.

Today, we take for granted that wherever we live, the public electricity supply will light our home, but this was not always the case.

chapter 3

An American Joke

Guides cannot master the subtleties of the American joke.

Mark Twain

Charles Batchellor was an Englishman who had joined Edison as an engineer fourteen years before the founding of the Edison Electric Light Company. The two men had first worked together on the telegraphic relay that Edison had manufactured in Newark. Later, Batchellor had shared in the success of Edison's phonograph, and over the years had become a close friend and trusted advisor. He was a tall, dark, thick-set man whom women thought quite handsome in his younger days. When Tesla met him, his hair and beard were still thick and black, even though he was slightly older than Edison.

Edison had a company in Paris which was having

problems, so he sent Batchellor over to sort them out, telling him to stop the Paris engineers from making so many mistakes and get them earning money. To be fair, the firm was working in a new area with unskilled engineers and untried equipment. Nobody quite knew what they were supposed to be doing and most of the mistakes were made by well-meaning staff doing their ignorant incompetent best.

The Continental Edison Company of Paris was a subsidiary of the Edison Electric Light Company that made dynamos, motors and lighting systems. However, not all its sales had been successful, certainly not the complete lighting system and power station that it had supplied to a German railway. Unfortunately the staff who installed this at Strasbourg station made some serious mistakes, and, as the new lighting system was being opened by Emperor William I, it short-circuited, exploded and demolished a wall right in front of the Emperor and his party. Neither the customer nor the Emperor were impressed, especially as, in 1883, not long after the end of the Franco-Prussian War, there was still quite lot of tension and ill-feeling between France and Germany. A French company almost killing a German emperor by installing an unsafe electrical system was not looked on kindly by the German authorities, and this 'accident' by Continental Edison was looked on with great suspicion.

The situation was extremely embarrassing for Edison and looked as if it could destroy his reputation in Europe. There was also a danger that the company would not be paid, and might even be sued for damages. Certainly every step taken to correct the installation would need approval in writing by the Germans; and so that they could be sure that the installation would never be dangerous again, Edison sent his most trusted lieutenant, Charles Batchellor, across to Paris to ensure that the problem was solved quickly.

At this time, Nikola Tesla had not been working for long with Continental Edison. After graduating from Prague University he had first worked for the telephone company in Budapest until it was sold, and then moved with his boss to Paris. Tesla hoped that this move would give him the chance to develop his ideas for AC electric power. When Batchellor arrived in Paris, he was just settling in at Continental Edison.

Bubbling with ideas, he would talk endlessly to anyone who would listen. Although he had no appreciation of the need for commercial secrecy and little feel for how business worked, he could play a mean game of billiards. There were quite a few American engineers at Edison Continental who, being far from their home comforts, often played billiards in the evenings. Tesla would join them and, during the games, would talk freely about the ideas he had for making an alternating current motor and the bright future that he saw for alternating current.

The type of electricity that the Edison companies made was DC. Tesla had worked out a theory of AC electricity and believed that this would replace DC, once people realized its advantages. Few of the people he worked with in Paris, however, saw any market value in his ideas. All the Edison company's patents were for DC inventions, and all the kit they sold used DC. When a company has a monopoly on an invention, it is hardly likely to be interested in a new idea that will make it obsolete and, anyway, it had just made a large investment in DC power stations and wiring systems.

Likewise, the new DC science was causing enough practical problems for the few skilled engineers who appreciated its theory without scrapping everything to start afresh on some wild new idea. They saw Tesla as a pleasant, talented, but slightly mad man, and put him to work on something practical – building direct current motors and working out how to couple dynamos together.

One of Tesla's billiard partners, a Mr D. Cunningham, was the foreman of the Mechanical Department. He must have believed in what Tesla was saying because he made the young engineer a serious business offer. He suggested setting up a joint stock company to develop Tesla's AC ideas. Cunningham knew that this was how Edison had begun in New York and that he had succeeded in luring lots of money out of the rich telegraph companies. The whole course of Tesla's life and fortune would have been very different if he had understood what Cunningham was offering him, but, instead, he burst out laughing at the very idea. Offended, Cunningham thought better of trying to work with such an unworldly dreamer, so the first of Tesla's many chances to make his fortune was thrown away.

Despite his business naiveté Tesla was an extremely

competent engineer. He set to work on the company's power plants where his talent quickly showed in the improvements he made to the designs of the DC motors and the automatic dynamo regulator he invented. When Batchellor came to Paris he quickly recognized Tesla's potential and made him the company's chief trouble-shooter.

Desperate measures were needed to sort out the disastrous Strasbourg contract. And, because in addition to being practical, Tesla also understood theory, and was the best engineer in Paris, he was promptly sent to get the system working. He had already been half-promised a bonus for his work on the dynamo regulator, but no money had been paid. Now Tesla's manager hinted that if the Strasbourg system was put right quickly, a substantial bonus of about $25,000 might be paid.

The technical problems of the system were not difficult for Tesla to solve, but the Germans, no doubt still suspicious and wanting to be sure that there would be no more explosions, insisted on written approval for every small step he made. The bureaucratic processes left Tesla with lots of spare time, hanging around for permission to act. He didn't waste this time, however; he rented a small backstreet workshop and built the world's first AC motor and a small two-phase AC generator to power the induction motor he had dreamed of for so long.

This AC motor was far more efficient than any other motor at that time, simpler to build and wouldn't need to have its commutator brushes replaced all the time.

Before this, nobody had managed to make any sort of AC motor because when alternating current was applied to an ordinary DC motor, it just made the motor vibrate without turning it. Tesla, however, had the inspired idea to use more than one AC supply. By making the two supplies work together he was able to make a magnetic field which rotated and carried the shaft of the motor around with it, without inefficient brushes to connect to the moving parts. Not only had he got rid of the sparking, wasteful brushes of the DC motor, he had created an engine which would run from the versatile AC electricity he was so enthusiastic about.

So, soon after his twenty-seventh birthday, Tesla ran his first small-scale AC motor in a backstreet of Strasbourg.

Now, with a prototype which proved that his idea really worked, and having turned down Cunningham's idea for forming a joint stock company, he needed money for more research. But, sadly, once again he showed the complete lack of business sense that would dog him for the rest of his life. Being easily impressed by social position and civic pomp, he had struck up a friendship with a former mayor of the city. Confusing civic status with business sense, he invited ex-Mayor Bauzin and his friends from the city council to see his new motor, hoping that the rich burghers of Strasbourg would give him financial support to build AC motors. But they were incapable of understanding the possibilities of his invention. All they saw was yet another generator and motor that went round just like the Edison motors, and having just invested lots of time and money in an Edison power station, they saw no reason to scrap it for another.

If Tesla was disappointed at their lack of enthusiasm he tried not to show it. His friend, the ex-mayor, using all his political skill, assured Tesla that his motor was a tremendous success and that if he went back to Paris, he would be sure of a triumphant reception. Tesla, believing him, resolved that when he got the promised $25,000 he would use it to build a full-size motor and generator. Once he could show this working, he was sure that everybody would want to switch to his new system and scrap all their existing investments.

Returning to Paris, Tesla went straight to the operations manager who had sent him to Strasbourg. He explained how he had restored both the lighting system and the company's reputation, then asked for the promised bonus. The manager, although overwhelming in his praise and thanks, explained that any additional payment would have to be approved by the accountant. Off Tesla went to the accountant, who again was profuse with the company's thanks, but regretfully explained that additional payments would need the approval of the chief executive. Undeterred, Tesla made an appointment with the chief executive who, again unstinting with his praise and thanks, explained that additional payments were a matter that he left to the operations manager concerned and he never interfered in the decisions of his operations managers. It didn't take any more trips around this fruitless ring to convince even the optimistic Tesla that he was not going to see any cash.

As far as the company was concerned, the work had been done, the payment had been banked, and the company's reputation had been saved. If something else went wrong, it could always find another young engineer to work day and night to put things right. Funny how engineers always expected more money even when company rules didn't encourage additional payments.

Perhaps this summary is unkind. Perhaps the company simply did not have the money to spare; perhaps it was all a 'misunderstanding'. Perhaps it was not; perhaps the company simply exploited Tesla.

Like many another young engineer, Tesla had just been given his first lesson in protecting self-interest: To avoid 'misunderstandings', get offers in writing *before* you do the job. But, as Tesla confided to Charles Batchellor, his friend and regular billiard partner, 'I expected the company to keep its word.'

Did Batchellor play a role in Tesla's disappointment? He was, after all, a close personal friend and commercial advisor to Edison, and his loyalty lay with that man. Over many a Paris billiard match, he'd heard Tesla's comments on the future of electrical power and had seen both potential and threat in the young engineer's work. The letter of introduction he had written for Tesla to Edison showed that he took the young engineer seriously. One sentence said: 'I know two great men and you are one of them; the other is this young man.' Batchellor could well have been worried that Tesla's ideas might topple Edison's growing electrical empire, and could from the start have reasoned that it was better to have a cash-strapped potential rival working for you.

When Batchellor persuaded Tesla to go to the United States to work for Edison, he was fully aware of the difficulties that Edison was having in getting his ideas to work reliably in the real world. A practical man who neglected theory, Edison developed things by trial and error. Of late, he'd had a run of errors! In Tesla's mathematical approach to engineering, Batchellor saw a way of avoiding the cost of too many random tests. Tesla could work out the consequences of his engineering actions before he tried them. Perhaps, in suggesting that Tesla went to the United States, Batchellor hoped to harness Tesla's foresight for Edison's benefit and to use the young engineer to help the American company solve its problems of implementation. The youngster was obviously talented. But we may never know if Batchellor was taking advantage of Tesla's ill-treatment by the Paris management or if he contrived it. Either way, the result was the same: Tesla was persuaded to sell his books and personal effects, scrap his models, pack his designs in a travelling trunk and emigrate to America.

It does seem likely that Batchellor's last job in Paris was to persuade Tesla to go to the aid of Edison's American operation because he certainly lost no time in returning to New York himself. The rest is history. All in all, it's hard to avoid the suspicion that Batchellor set up the ill-treatment of Tesla in the hope of pushing him off to America.

Before Tesla arrived in New York, the biggest problem Edison had was connecting dynamos together. DC generators create short pulses of current which get stronger and weaker in time with the speed of the rotating coils of the generator. Because the motors and the generators are exactly the same machine, if one generator is at the peak of a pulse output when the other it is connected to is at the bottom of its cycle, the high one will try and drive the low one as a motor. Unless both generators are set so that they rotate at exactly the same speed, this coupling effect will cause a see-saw-like pulling and pushing as each generator tries in turn to drive the other as a motor.

Edison's early steam-driven dynamos could only drive a small number of lights, not enough to light all the lamps that his customers wanted to connect to his public system. At first he had simply tried to make a bigger generator, which, because he was a great fan of Barnum's Circus, he had named 'Jumbo', after one of Barnum's performing elephants. But even a 'Jumbo' generator could only light up about 400 lamps, and that was not enough to run a public electricity supply.

Edison could not make a single generator big enough to serve all the customers for his public lighting system. He had to couple two or more dynamos together to supply enough power, but as he didn't understand the synchronization of current pulses, his coupled generators did not work as he expected.

The first time Edison connected two of his 'Jumbo' generators together he realized that he had a serious problem. Speaking later of that test he said: 'We started the other Jumbo and threw them in parallel. Of all the circuses since Adam was born we had the worst then.'

Edison made a mechanical link between the steam valves of the two generators to stop the 'hunting' caused by the unsynchronized pulses of current. His jury-rigged arrangement allowed that particular pair of generators to work together, but Edison had not tackled the electrical control solution needed to make any generator run with any other generator. John Hopkinson, one of his scientific advisors, wrote to him about the flaws in his dynamos, saying, 'It is necessary to make a critical study of the generators with a view to not only improving them but of placing ourselves in a position to say beforehand how we should modify the machines to meet varying conditions.'

Edison, however, who disliked applying mathematical theories to his work, preferred to have a go at making a thing and then fiddling with it in the hope of sorting out all the 'bugs'. Sometimes this approach worked, but the failure of his non-analytical approach was causing a lot of embarrassment.

Long before he had left for Paris, Batchellor had been aware of Edison's problems with coupling his dynamos together. Indeed he had been involved in suggesting a solution for some of the larger private systems. For example, Haverly's Theatre in Chicago had an Edison Lighting System that ran 647 lights. This was far more than a single dynamo could power and was more than half as many again as the 400 lamps that a 'Jumbo' could light. The Edison Electric Light Company Bulletin reported this impressive performance, but never explained how it was done. The theatre installation was really three separate lighting systems in the same building. One dynamo powered the lights in the vestibules and entrances, another the auditorium lights, another the stage and dressing-rooms lights. The three systems were not connected together.

Batchellor had realized that Tesla was capable of solving the coupling problem when the young engineer had designed an automatic dynamo regulator – and Edison, in his turn, knew how to motivate Nikola. Realizing that Tesla was hungry for capital to research his AC ideas and that he would be prepared to work long and hard for a chance to develop his inventions, Edison offered Tesla $50,000 to improve the designs of the generators for his central power stations.

Tesla, not having learned any business sense from his time in Paris, set to work with a will, pushing himself for the next year from 10 a.m. to 5 a.m. the following morning, never taking time off. He produced new designs for twenty-four different types of standard DC machines; designs which not only gave much higher currents but were easy to regulate and couple together. He also designed a control system which made sure that when the generators were connected together the current pulses they generated were always synchronized. This meant that the generators never tried to work against each other, the regulator ensuring they always worked together. Later, Tesla recalled what Edison said to him when he solved the coupling problem: 'I've had many hard-working assistants, but you take the cake.'

Edison built Tesla's new generators and tested them out. Tesla's 'third brush regulation', which added an extra pair of brushes to Edison's design, enabled any number of dynamos to be connected together. The machines worked well, so Edison took out patents and started to replace his existing dynamos. At this point, Tesla asked for the $50,000 he had been promised. At last, he thought, he would have enough money to build a full-size AC motor. The long hours of work had paid off and he would now have enough capital to get started. Then, for the second time in two years, Tesla was cheated by an Edison company.

Going into Edison's office, he reminded him of the $50,000 he had been promised for improving the design of the generators. Edison's reply upset Tesla so much that he quoted it many times afterwards. Edison said, 'Tesla, you don't understand our American humour.'

Once again, Tesla had been taken for a sucker by a hard-nosed Edison company. The first time it happened he must have consoled himself with the thought that Edison could not have known what was happening. Indeed, why else had he been persuaded to travel halfway across the world, losing all his possessions on the trip, just to work eighteen hours a day, seven days a week for Edison. He even gave Edison the benefit of the doubt over his early hostility to AC electricity. In a frenzy of hero-worship, he had set out to impress Edison with his worth, showing how the flawed DC machines that Edison had designed could be improved and, in the process, ensuring the early commercial success of Edison's DC system.

Edison, though, was not interested in Tesla's devotion, viewing him as a savage barbarian who happened to be good with motors. At that time Edison had never been to Europe and held strange views about it being a backward barbaric place. He was once overheard asking Tesla if he had ever eaten human flesh, possibly imagining that Croatia was part of an uncivilized jungle in central Europe. But Tesla was no savage barbarian, and the flagrant breaking of the verbal promise finally shocked his eyes open. He had been brought up as a European gentleman, within a family which for generations had served either the church or the army. As far as he was concerned, a gentleman's word was a binding obligation.

He resigned on the spot, giving no thought to his personal position. A year after his arrival in America, he was again out of a job. This time, however, he had the status of a good electrical engineer who had gained sound experience working for the respected Edison Electric Light Company.

chapter 4

Let There Be Light

Yea, the darkness hideth not from thee; but the night shineth as the day: the darkness and the light are both alike to thee.

Psalm 139

Making light after the sun has set is an age-old problem. When early man wanted to see in the dark he had to build himself a fire, but the fire was not a portable light unless he broke a branch off a tree, and lit it from the fire. Even then, the crude torch had a limited life and didn't cast much light. Some time in early history, somebody came up with the idea of making a stone or pottery bowl, filling it with animal fat or oil and floating a burning rush-wick in it. The wick would draw up oil and continue to burn until the oil ran out. These lamps were portable and burned for longer than a crude branch torch. By biblical times, these rush-lights were in common use.

The candle, a tube of slow-burning material, such as tallow or wax, with a fibrous wick running through the middle of it, was another portable solution to the problem of creating light. Candles were used as early as 3000 BC and became a very popular form of domestic lighting because they are simple to make and use, and give good light.

The next step forward in lighting was made in the Yorkshire town of Wakefield, in the North of England. In 1684, the vicar of that town was the Revd John Clayton, an amateur scientist when he wasn't attending to his parishioners' souls. Interested in the coal mined in Yorkshire, he knew that coal could give off an explosive gas. Having heated some coal, he discovered that it gave off a lot of gas that would burn, and that, after the gas had been driven off from the coal, a useful tar was left behind.

Making tar involves heating up coal without letting it burn away. To stop the coal simply flaming up, it has to be heated inside a closed iron container using a process that scientists call distillation.

A century later William Murdock, a Scot from Ayrshire, realized that if he heated coal in a closed container and piped the gas off it, that gas could be used to do other things. This gave him the idea of burning the 'coal gas' he produced to make light. In 1792, he built a small coal distillation plant behind his house in Redruth, Cornwall, and made enough gas to light his own house. The fame of his lighting system spread and he was approached by a number of industrialists to make similar lighting systems for their factories. Within the next two years, he had built successful gas-lighting plants at the factories of Boulton and Watt in Birmingham and for the cotton mill of Phillips and Lee in Manchester.

Murdock used this burning coal gas system to light over 1,000 gas lights in the Phillips and Lee cotton mill, thus making it easier and safer for the machinery to be worked after dark. To make the gas for lighting, coal has to be heated in a special sealed container called a retort. A cotton mill or a factory could afford such equipment, but it was too big and costly for lighting individual homes. Gas lighting, however, is much simpler to use than candles or oil-lamps because the fuel for the light can be supplied down a pipe, and as long as the gas flows, the light will stay lit. It will not burn out like a candle or run dry like an oil-lamp. Murdock's fellow scientists were very impressed with his work in developing gas lighting, and, in 1808, awarded him the Rumford Gold Medal of the Royal Society for the Advancement of Science.

With the discovery of gas lighting, which solved the problem of maintaining a fuel supply for lights, the lighting of public places became possible, and soon became a popular way of lighting towns. The first street to be lit by the new gas lights was Pall Mall, London, soon to be followed by Westminster Bridge and the streets of Westminster. Because the street lamps had to be turned on and lit at night and then turned off again at dawn, a lamp-lighter − a man with a pole and lighter − was employed to do this work. Twice a day, at dusk and dawn, he had to visit each lamp to light it and then put it out again.

The lighting companies made it easier for the lamp-lighter to do his work by using something called a pilot light, a small flame that burned day and night. To turn the lamp on, all the lamp-lighter had to do was pull open a valve to let the gas flow. The pilot light then lit the lamp flame.

Once towns built municipal gas works and laid down networks of gas pipes, it became easy for people to light their homes with burning gas. By 1875, most town houses were lit by gas. The gas was burnt in unshielded flat-flame burners, known as batwing burners because they were shaped like a bat's wing. One ton of coal would make 16,000 cubic feet of gas and, by 1855, the gas industry was supplying over a million cubic feet of gas per day to London alone. Vast gasometers, the large steel cylinders needed to store gas to meet the needs of the town, started to appear on city skylines.

A German inventor, Karl Auer von Welsbach, improved the amount of light given off by gas lights when he made what he called a gas mantle. This is a thin wire globe that fits on the burning gas and is coated with chemicals to make the flame glow more brightly. His mantles were used in street lighting and would give as much light as 4,000 candles. But mantles were fragile and easily damaged when lighting the gas.

There was a crude type of electric light, known as arc lighting, available in the mid-nineteenth century. This was first used in the South Foreland lighthouse in 1858.

The light from an arc lamp was powerful and worked by creating a continuous spark between two carbon rods. The light came from this controlled streak of 'lightning', but it needed constant attention from skilled engineers to keep it working. Arc lamps worked at very high currents and very low voltages. To take the high current from the generator to the arc lamp, a very thick copper wire had to be used which made these lamps expensive. Unless a more reliable arc lamp could be made, it would not be suitable for unattended use as street lighting.

The world was becoming a lighter place, but the chore of lighting up and turning off each separate gas lamp, one at a time, still had to be carried out every evening and morning. Coal gas was also highly poisonous. If you did not notice that your light had blown out, the gas would continue to flow and kill you, often quite quickly. There was also the danger of an explosion if gas from the supply pipes leaked into a building. Modern gas suppliers use natural gas that is not poisonous, but great care still has to be taken to avoid gas leaks and their explosive results.

Electric arc lights worked by slowly burning away carbon electrodes to make a continuous spark, but unless the spark-gap was constantly adjusted the light stopped working. Arc lamps were used in lighthouses, where each light had a keeper to tend it, but were not practical for lighting people's homes or public streets. Gas lamps only needed attention when being turned on and off.

Thomas Edison realized that if he wanted to create an electric lighting business, he would have to supply a better product than the gas lighting companies. He knew that a British scientist, Sir Joseph Swan, had made an incandescent light bulb (a light bulb that glowed like an eternal flame). But Swan's flame, far from being eternal, only lasted for a few minutes before burning away. He used a strip of carbon-covered paper inside a glass bulb from which all the air had been removed and passed an electric current through the strip.

The strip would glow white-hot and give off light, but what was more important was that it didn't need to be lit by a match and could be turned on and off from a switch placed near the door as you entered. Edison was a genius at taking other people's under-developed ideas and turning them into something practical. This is what he did with Swan's lamp. He tried different materials, different types of glass bulb and different gases in the bulbs, and succeeded in making an incandescent light bulb that worked. His first light bulb lasted for forty hours, not long compared to our modern light bulb's 1,000 hours' lifespan, but a remarkable feat for 1880. Astute business man that he was, Edison knew that it was not enough just to make a successful lamp, he had to offer a complete service. If he wanted to sell light bulbs, he had to supply electricity as well.

In 1882, Edison opened his first electric power station in Pearl Street, New York. It supplied eighty customers with sufficient electric power to light 400 light bulbs. For the very first time, thanks to Edison's new DC electricity system, it was possible to dispel darkness with the flick of a switch. The downside, however, was that light could only come from the flick of a switch in a few privileged places because Edison was only able to transmit his DC electricity for about half a mile from his power station. So, if anybody wanted electric lighting they had to be careful where they lived or be able to afford to set up their own power station. Outside New York, the bulk of American homes continued to be lit by either gas, oil lamp or candles; and very few public places had any lighting at all.

In 1885, because Edison had never taken any interest in public lighting, he did not control the entire American electricity industry. Therefore, when Nikola Tesla left Edison's company he was able to apply his engineering skill to an area of lighting that was ripe for development, and, what's more, there were people who were prepared to fund him.

During his year in America, Tesla had learned a little about business, but not as much as he thought he had. His friend, Ernest Osborne, introduced him to a group of businessmen who suggested forming a joint stock company to produce arc-lamps for public and industrial lighting. Tesla, who had still not learned the lesson that it was too late to start negotiating after signing up, was about to learn the hard way.

Tesla realized that Edison's ability to control his own research had come from his position within a joint stock company. He knew that such a company provided money for experiments and earned it back by selling the resulting inventions. What he had not grasped was that the objectives of a company are controlled by its founders; and these founders offered him only a small salary, a few shares in the company and a laboratory in Liberty Street, New York, to produce arc-lights. His shareholding was too small to give him any control and he was not offered any say in the running of the company, but he entered into the venture believing that he would be able to convince his majority shareholders to support his AC work. He was wrong.

Speaking of it later, he recalled, 'Here, finally, was an opportunity to develop the motor, but when I broached the subject to my new associates they said, "No, we want the arc lamp. We don't care for this alternating current of yours."'

Tesla had not realized what a useless invention his AC motor was without a public AC power supply to drive it. If he wanted to sell AC motors he would also have to create a complete public AC electricity supply infrastructure: power stations, distribution cables, junction boxes, sub-stations, transformers and household metering systems would be needed. His business partners knew this and realized it was way beyond their capital resources to produce such a system. The only public power supply was Edison's DC, one that could not run Tesla's motor, even if he could make it.

The men of business saw no money in a motor that couldn't be plugged into the only power supply in town. Money could be made lighting cities at night, not in making a motor nobody wanted. City lighting needed much more powerful lights than the puny carbon filament lamps that Edison was making his money from. His incandescent lamp had the power of about sixteen candles, but this was not enough light to brighten the public spaces of the American cities. The arc lamp was a much brighter source of light and Edison was not a competitor in this unexploited market. Here was a clear gap to be filled by enterprising businessmen, provided they could make a good arc lamp. Tesla could make the arc lamp they needed and he did.

The arc lamp gives a bright light, but its light shines only as long as the carbon rods last. The electricity burns away the carbon until the gap is too wide for the spark to jump, the current stops flowing, and the light goes out.

The first practical arc lamps were made by a Russian, called Paul Jablochkoff, about ten years earlier. He had been the director of the telegraph system connecting Moscow and Kursk until 1875, when he decided to give up his job and travel to Philadelphia to go to the Great Exhibition of 1876. He only got as far as Paris, however, because he stopped off to work for a while as an engineer to replenish his funds. During this time, he invented the first practical arc lamp and was given a contract by the city fathers to light some of the public places.

The Jablochkoff candle, as his invention became known, had two parallel carbon rods separated by a thin layer of plaster of paris. The lower ends of the rods fitted into short brass tubes that went into a holder to make a connection with the power source. A thin strip of carbon connected the two rods at the upper end. When the candle was switched on, the thin track of carbon burned out, starting the arc. As the carbon rods burnt away the plaster crumbled exposing fresh carbon. This would continue until all the carbon had burnt away. A Jablochkoff candle would only last for about ninety minutes when it would need to be replaced with a new candle and it could only be lit once. If the current was turned off, the candle would not relight.

When Tesla lived in Paris, he had often walked at night along the brightly lit Avenue de l'Opéra into the Place de l'Opéra. He had seen the forty-six lamp columns each bearing a 'Chandelier' that changed candles mechanically as they burnt out. He was quite sure he could improve on Jablochkoff's simple design and make a better arc-light. He was also sure that if he designed and built a better arc-light, his fellow shareholders would then support the development of his AC motor. For almost two years, as well as designing a self-starting arc-lamp that had an automatic feed mechanism for continuously replacing the carbon rods as they burnt away, Tesla used his own small salary and spare time to work with AC generators and motors. He then went back to the directors of the now prospering company and again asked for their support to develop his AC motor ideas. Once again, however, he was told that they were not interested. The company had been set up to make and sell arc-lights and that was what it was doing. If Mr Tesla wanted to work on his own, then he was free to do so.

Tesla asked how much his shares in the company were worth, and learned a painful lesson about private joint stock companies. As the shares were not quoted on the stock market, he could only sell his share to a buyer approved by the majority of the stockholders and no other stockholder wanted to buy his shares. He made the bitter comment, 'In

1886, my system of arc lighting was perfected and adopted for factory and municipal lighting, and I was free, but with no other possession than a beautifully engraved certificate of stock of hypothetical value.'

Again, Tesla walked away from men he was too polite to call dishonourable, but whom he neither understood nor could work with any longer. This time there was no job-offer forthcoming for the brilliant engineer who was getting a reputation for being difficult to work with. He still had to eat, though, so his revolutionary motor sat in his boarding house while he spent his days heaving a shovel. His long education and superb engineering skill had not brought him fame and fortune; and his workaholicism had only helped his rivals. His obsession with AC electricity had pushed him out of what others would have seen as a promising career into a labourer's job. The Serbian dreamer's visions sat uncomfortably alongside the harsh realities of nineteenth-century business practices.

At that moment, inventions that we all take for granted and rely on today looked as if they were never going to happen.

Tesla had understood that the problem with Edison's system was that there is a limit to the voltage his direct current generators could produce. For all practical purposes, this was about 115 volts. Edison's system could not run at the high voltages we use today, so, as mentioned before, it ran out of pressure about half a mile from the power station. For rich people this was not a difficulty, they could afford to build their own power station close to, or in, their houses. People who couldn't afford to do that, couldn't have electric power. Edison's early success had relied on selling 'isolated' lighting sets to wealthy people. His system was never going to provide the cheap power that Tesla wanted to achieve for everybody.

Tesla, who understood Ohm's law, applied it creatively. He had discovered that he could transform the voltage of his AC electricity, changing it upwards or downwards using two linked coils of wire called a 'transformer'. He could transmit this power at high voltage and low current over long thin wires. When he wanted to use the power, he could turn it back to low voltage and high current. He was using Ohm's law to full advantage, but was having tremendous problems in persuading anybody to listen to him. Having a wonderful idea, he had discovered, didn't mean that it would be accepted.

Today, we are so used to having electric power anywhere we want it, that it is hard to imagine life without it; and we can be forgiven for believing that nobody would resist or argue against such a helpful invention. But, at that time, Tesla's electricity system challenged established inventors who had invested interests in less efficient systems. This didn't make Tesla popular. His rivals realized he had invented a system that could destroy their business investments. Tesla, however, driven by his need to prove himself the 'worthy' inventor his parents could be proud of, and the conviction that he was on the right track, was not going to give up his dream of universal AC power or prostitute his skills making arc lamps for money. He would rather scrape a living digging ditches.

chapter 5

An Unreasonable Man?

*George Bernard Shaw once observed that all progress
depends on the unreasonable man. His argument was
that the reasonable man adapts himself to the world
while the unreasonable persists in trying to adapt the
world to himself, therefore for any change of consequence we must look to the unreasonable man . . .*

Charles Handy

Successful businessmen are short-term thinkers. If they
fail to meet the payroll at the end of the week, they stop
being businessmen and become bankrupts. This harsh
discipline forces a businessman to ask key questions:
How much will it cost to make? How many people will
want to buy it? How soon can I sell it, and how quickly
will I get paid?

Some products are easy to make and sell. The candle

was – and is – a successful product because it is simple to make and sell, and requires no after-sales service. This is good business. The manufacturer makes and supplies one product, the customer comes back for more. Almost the whole of the grocery and hardware trade was made up of products of this type. American business pioneers understood the need for basic foodstuffs and simple tools, and the great Sears and Roebuck empire was built selling goods like this.

As society became more sophisticated and complex, however, new types of product were needed. The simple candle was replaced by gas light, which was much more convenient for the customer to use but more complicated for the businessman because the customer can only use the gas light if she has a gas supply. Therefore, if he wanted to sell gas lights he must first provide a gas supply system and the gas, and the gas must be on tap twenty-four hours a day.

To make a gas light work, then, requires a whole string of events. First, a coke oven must be built to heat the coal and a large tank installed to store the gas that is made. Underground pipes must be laid to connect the gas pipes to customers' houses. For this venture, the businessman has to find the money to hire workmen to build the gas works, lay the pipes and keep the gas flowing day and night. Gas meters also have to be read, bills sent out and vital ongoing maintenance work, such as the fixing of gas leaks, carried out. This capital expense has to be risked long before customers even consider buying a gas lamp, and such investment can take years to recover before the businessman can make a profit on his initial outlay. If, during this time, somebody comes up with a new and better way of providing light, the investment is a financial disaster.

Now, in 1878, Edison, who had started out with nothing, lived by his wits, but had shown a remarkable understanding of people's needs when he set out to develop his electric lighting system, had come along with a new, more convenient way of making light which posed an enormous threat to the gas lighting industry.

Edison's knowledge of science might have been sketchy, but he was a great showman who had considerable skill when it came to marketing. As a result, he had succeeded in ensuring that people wanted his electric light, even before he had built his first power station. In fact, although Edison is remembered as the 'Great Inventor', he was also the first master of 'hype'. To back up his publicity displays – the parades of electrically-lit showgirls – he had also employed a sales force to sell his systems, and motivated these men by posting them the Edison Electric Light Company Bulletin every ten days. Alongside enthusiastic reports of new installations, the Bulletin never missed an opportunity to run down the competition's gas lights. The following snippet, describing an Edison Electric Light Company installation at a wholesale grocers, is typical of this approach:

In this room fifty clerks do clerical work all day. The heat from gas has proved injurious to health, and the gas light has proved injurious to eyesight. This room

*is now lighted by one of our isolated plants and the
injurious effects of gas are entirely removed.*

The more exciting gas explosions were guaranteed
explicit lurid write-ups; and the inaccurate finding that
gas light caused short-sightedness was given a banner
headline.

So, while Nikola Tesla had been busy fixing a dynamo
in a back street on his first day in New York, Edison
had been fully stretched in his battle with the rich
well entrenched gas-light industry. And the contents
of his Bulletin showed that he had no scruples about
spreading misleading statements concerning his competitors. Even *The Operator and Electrical World*, one of
the leading electrical journals of the time, said, tongue
in cheek, of Edison: 'We quote from a professedly
scientific report, which adds that the great succes-
sor of Barnum realises 70 per cent of the original
energy and apparently hopes to double it!' (The Barnum
referred to in this comment was P.T. Barnum, a famous
American Circus showman of the time. Obviously,
Edison's marketing showmanship exceeded his grasp of
arithmetic!)

He made sure that his customers knew that his electric
light was much better than gas because he told them
and kept on telling them. He made sure all the faults
of gas light were repeated over and over to potential
customers. Gas was dangerous; it caused explosions
when it leaked and, when it leaked, Edison made sure
that all his salesmen knew and told.

When he set out to replace the rich and successful gas-lighting industry, Edison designed a complete alternative system from power station to fancy lampshade, and sold the American public the idea of a safer, cleaner, healthier, cheaper, more convenient and brighter light for them and their loved ones.

Any change challenges somebody else's wealth, prestige and power. The gas companies had displaced the candle makers and oil-lamp vendors; now Edison challenged them for the ground they had taken.

Although he had been successful in everything he had done thus far, replacing the gas lighting industry with an electric lighting industry would require more money than he had. Having become very friendly with Grosvenor Lowrey, one of New York's leading patent lawyers, he decided to discuss his ideas and need for financial back-up with him. Lowrey, appreciating the potential, decided to try to form a group of capitalists to provide money for Edison, and set about talking to the directors of the powerful Western Union Telegraph Co., and to other rich clients.

They, in turn, set up the Edison Electric Light Company with an initial capital of $300,000 made up of 3,000 shares; 2,500 were given to Edison, and the other 500 were bought by the syndicate for $50,000. In return, Edison assigned to the company all the electric lighting patents he intended to invent for the next five years. In effect, the investors had formed a company to exploit an invention that did not exist. Today, we would call this a venture capital company, but in 1878 it was a first.

Edison himself commented, 'Their money was invested in confidence of my ability to bring it back again.'

The main man behind the capitalists who supported this venture was the banker JP Morgan. Morgan played an important part in the development of electricity in America and will also figure in the fortunes of Nikola Tesla later in this story.

Having offered his financial supporters a chance to seize the wealth and prestige of the gas industry for themselves, Edison wooed the gas company's customers with visions of a better future. To do so, he preyed on their fears by spelling out the dangers that gas presented. 'Do you want your wife and children to go blind, to be burned to death in explosions?' In this way, he made the possibility of electric light seem not just a new product, but a deliverance from evil, and became Moses leading his customers to the Promised Land.

One reason why Edison had been so hostile towards Tesla's schemes to build an AC electricity system, was that Edison's company was in severe difficulty because of the restrictions of DC power. It had not matched up to the sales pitch he had made for it and he didn't want his backers to find out its limitations. Edison, in fact, was fighting for his financial life. He had built the first public electric power station in Pearl Street, New York, but it had cost far more than expected, taken much longer to construct, and had not attracted as many customers as planned. If the Edison electricity system was to succeed, more power stations were needed. Electrical fittings would have to be mass-produced to reduce their cost, but Edison's backers were refusing to risk more money manufacturing the fittings, and the lack of essential funding was stopping Edison developing his public electricity supply. To get round the problem, he had set up his own company to make fittings.

Three years earlier Edison had withdrawn all $78,000 profits from his previous inventions and sold his share of the Edison Electric Light company to his backers. Still he didn't have enough money to set up a factory to make generators and electric fittings. Forced to mortgage his future European earnings to borrow, it was small wonder that he had been so worried about the explosion of one of his systems in Strasbourg putting those earnings at risk. With his personal fortune entirely sunk into the grimy back streets of New York, if the factory he had set up at 104-106 Goerck Street failed Edison would be bankrupt.

Although he continued to show a brave successful face to the world, his business came very close to failing during the early 1880s. Meeting the weekly payroll became a regular worry and more than once he asked his company secretary, Sammy Insull: 'Sammy, do you think you can earn a living again as a stenographer? If you do, I think I can earn my living as a telegraph operator, so that we can be sure of having something to eat.'

The gas companies were watching Edison's every move. If he lost the public's confidence, they would exploit any problem to discredit him. As he pulled the switch to start the single working generator at Pearl Street at 3 p.m., on a Monday afternoon in September

1882, he said, 'Success means world-wide adoption of my central-station plan. Failure means loss of money and prestige and the setting back of my enterprise.'

The electricity system Edison had designed had to work if he was to recover his fortune. Whereas a single gas company could supply the whole of New York, Edison needed to build power stations every mile or so. Unless he could increase the power output of his stations, which meant coupling more than two dynamos together, he would be limited both in the distance he could deliver power and in the number of lamps he could light. This basic flaw in his original design was preventing him from making any profits from the expensive equipment he had already built. The need for large numbers of local power stations was going to be even more expensive and would reduce the popularity of the Edison system. He had, in fact, met a difficulty that his experience with the telegraph system had not prepared him for. He had hoped to supply his lighting current up to 20 miles (32 km) from the generator, as this was the distance a telegraph current would travel, but it hadn't worked out that way.

When Tesla arrived in New York, the sales of Edison's isolated lighting systems were the Edison Company's only real income. These systems were separate generator sets sold to people or businesses to light up individual buildings. By selling a complete power station and lighting system to each customer, Edison had avoided a problem he had been unable to solve, that of linking generators together to produce more current. Because each site was self-contained, the question of coupling the generators never arose. Nevertheless, the failure to solve the coupling problem had prevented him from expanding his public electricity supplies and from earning money from his investment in supply lines.

In addition, as mentioned earlier, Edison had faced enormous claims for damages from the owners of the *SS Oregon* when one of his isolated systems was fitted in the ship without a thought about how to maintain or remove it for servicing. He was also at a very low ebb because his wife was seriously ill (she died soon afterwards) and because the only successful part of his company was now facing ruin.

The reputation and sales figures for his isolated systems had to grow or his whole venture would collapse, but in 1884 this reputation was at risk.

Having put years of his life and his entire fortune into commercializing his product, Edison had conned Tesla into using his theoretical and design skills to solve the fundamental problems which he himself could not, and had then discarded the Serb without a second thought. Edison functioned well with ruthless businessmen and was adept at answering their questions about money and return on capital. Where such matters were concerned, he was a reasonable man. In contrast, Tesla knew little about what made business people tick. He lived in a world of ideas; loved to build mental models of his inventions and to imagine them working. When asked, 'How much will it cost to make?' he would simply point to the elegance of the rotating magnetic field. When

AN UNREASONABLE MAN?

asked, 'How many people will want to buy it?' he would simply demonstrate how efficient it was. Asked, 'How soon can I sell it?' he would simply reply, 'Just give me a workshop and I will build you one.' 'How quickly will I get my money back?' was a question he was not even prepared to consider. He was working for the betterment of mankind – who could put a price on that? In short, Tesla was naive enough to believe that Edison would be prepared to throw away his life's work, junk the investment of his backers, and scrap everything he had done in order to start afresh with AC.

The possibility that if he did this, Edison could lose all the hard-won ground to the gas companies did not even occur to Tesla. He could only see that alternating current was far superior to direct current, and believed that Edison must see this, too. Little wonder that the practical Edison came to regard Tesla as a gullible fool; while Tesla came to see Edison as a loud-mouthed fraud who was trying fob off the public with an inferior power system.

Tesla's adventures in arc-lighting only served to prove to Edison that he was justified in using, then ditching the troublesome young Serb. The man, he decided, was so unreasonable that no sensible businessman could ever work with him. Reasonable inventors expected the holders of wealth and power to suppress or censor their ideas, and were prepared to court the favour of bankers to avoid conflict. After all, unless financiers believed an invention would increase their own wealth, power and prestige, they would not back it. Only a fool would tell backers they didn't know what they were doing; and only an unreasonable inventor would put himself out of work during an economic depression when jobs are so hard to come by.

By leaving Edison's employ, Tesla had ruled out working on DC systems and had also put himself out of the arc-lighting business. Being independent can be a virtue – and having faith in your own ideas is essential if you are an inventor – but inventors also need financial backing, laboratories and equipment. A shovel and pile of mud are not raw materials for building a world-wide power system!

Digging sewer ditches is physically exhausting, and trying to live on wages of two dollars a day must have made Tesla wonder why he was doing it; what he was trying to prove. He may not have made a fortune working for Edison, but he had had the security of a regular job that he enjoyed doing. Why did he persist in expecting businessmen to support his rotating magnetic gizmo; and why did he continue to follow his vision of transforming the power of flowing water into AC electricity to improve everyone's life? The answer was that he was still trying to prove himself to his dead father.

Always obsessed with personal cleanliness, Tesla was fanatical about washing. He would never use the same towel twice, always insisting on a freshly laundered cloth each time he washed. The squalor of his now daily working conditions must have been intolerable. He said of that time, 'I lived though a year of terrible heartaches and bitter tears, my suffering intensified by material want.' Surely he must have wondered if what he wanted was worth the personal cost. Perhaps he even considered going back to designing motors for Edison. Such moments, however, must have been few because he knew that his AC system was far better than anything else available. He just couldn't understand why nobody else seemed the least bit interested in it.

As he swung his pick, Tesla bemoaned his fate and, during his lunch break, would tell the foreman about his dreams, inventions and hopes for the future of electricity. The year, 1887, was a strange year. Many people were down on their luck and had to take whatever work was available. The foreman, who just happened to be a friend of Mr A.K. Brown of the Western Union Telegraph Company, repeated Tesla's dream to this man who was inspired by it. Brown had a steady job and an assured income. Perhaps he was also a bit of gambler who was able to appreciate that Tesla had ideas that could change the world. It was an outside chance, but one that he could afford if he could persuade a friend to share the gamble of financing the Serbian inventor long enough for him to produce some patents. The patents could then be sold and might make a profit. Brown was aware that Tesla had already made an extremely saleable arc-lamp and hoped that he might come up with something else that would sell. The bargain was struck; Nikola wiped the mud off his boots for the last time.

Having persuaded a friend to share the gamble, Brown formed the Tesla Electric Company. Tesla was given fifty per cent of the shares and a casting vote; Mr Brown and his friend fifty per cent, plus a half share in any inventions that Tesla succeeded in selling. They explained to Tesla that his motor would not be any use on its own; that he would also need to design generators, transformers and all the other bits and pieces that a full electricity supply system would require.

So, five years after his unsuccessful demonstration to the ex-mayor of Strasbourg and his friends, Tesla finally set to work on all the elements of his AC system. He built three complete working sets of AC motors that used different types of AC current. The simplest, which he called single-phase electricity, used two wires and the electricity reversed its direction sixty times a second. This is the type of current that is now used in most houses.

He also designed a two-phase system that used two linked currents and a three-phase that used three. The two-phase motor was the simplest to make, but the three-phase motor gave the most power for industrial use. Over the next six months Tesla invented over forty different types of motors, generators and transformers, and the transformers that were needed to change the voltage pressure of the electricity up or down. This ability to change the voltage made his system incredibly versatile. He could use thin wires to carry the electricity long distances at high voltages and low current, and a transformer could reduce the voltage and increase the current to drive the lights and motors at the site where the power was needed. After six months' labour, he

AN UNREASONABLE MAN?

was ready to demonstrate the equipment. He sent his two-phase motor to Cornell University where it was tested and found to be as efficient as the best of the direct-current motors of the time.

Tesla applied to the Patent Office for a single patent to cover his entire system with all its alternators, transformers and motors, but the office insisted that all his ideas should be broken down into a series of simpler and more detailed patents. His inventions were so novel that he was then granted thirty separate patents during the following year, which meant that he and the company now had complete commercial control of the as yet non-existent alternating current industry.

Such a high output of new patents attracted the attention of the engineering élite, and Tesla was greatly flattered when invited to give a lecture explaining his alternating current system to the American Institute of Electrical Engineers. Any aspiring engineer would have been pleased to speak to this august body, but for Tesla the invitation was proof that America's great and good were finally beginning to recognize him. This was the recognition he had dug ditches for; and at last he felt he had proved himself. His only disappointment was that his father had not lived to applaud his triumph.

The lecture was his big chance; he was going to be famous; social superiors would sit at his feet and marvel at the scientific elegance of his rotating magnetic fields. He had been right not to compromise. And, despite the fact that he was nearing the limit of Mr Brown's investment, and his patents had not yet earned a single dollar, he ignored all commercial concerns and threw his energy into preparing a lecture that would vindicate him in the eyes of the academic community.

The lecture he gave has become a classic of electrical engineering. He presented the theory of alternating current in power engineering; followed the maths with practical demonstrations of working machines; and showed that his group of patents were the foundation for the entire electrical system for the future. It was the most important step forward in electrical engineering theory ever to occur. With the benefit that hindsight gives, modern engineers now rank Tesla above Faraday as the father of modern electricity.

Tesla came out of the lecture on a 'high'. Speaking to Mr Brown afterwards, he said, 'It was a great pleasure to bring before the American Institute of Electrical Engineers the results of my work on alternate current motors. They received my ideas with great interest and I elicited considerable comment. With truly American generosity they bestowed a great deal of praise on me.'

Before the lecture, Mr Brown had been worrying about the financial aspects of his stake in Tesla. The cost of developing the prototypes and pursuing the forty patents had used up most of the company's money. These patents still had to be turned into products that could be sold, and Tesla was obviously not the man to do this. Fortunately, through his work in the Western Union Telegraph company, Brown had met an entrepreneur who wanted to challenge Edison's grip on public electricity systems, a man called George Westinghouse.

AN UNREASONABLE MAN?

George Westinghouse had been born into a family of railway engineers. After graduating from Schenectady University, he had worked in his father's successful rolling-stock factory in New York State. While an undergraduate, he had invented a gadget that allowed trains to change tracks, and had called this invention a 'railway frog' because it made a train 'jump' from track to track. With second-generation railway money behind him and a good engineering degree, he subsequently proved his commercial skill by inventing the air brake, a device that automatically stops a train if part of it becomes detached, and had sold this system to the railway companies for a serious sum of money.

If Brown was going to collect on his Tesla stake, he needed somebody to market Tesla's ideas and persuade the public to buy them. George Westinghouse, he decided, was just the man to do this – and he invited him to Tesla's lecture on 16 May 1888.

Westinghouse already knew a little about alternating current and Ohm's law. In 1883 he had bought the American rights to an English patent for an alternating current transformer. He knew that the $50,000 he had paid for this was money well spent. If he could use the cost advantages of high-voltage transmission and reduce the voltage pressure to a safe level in people's homes, he knew he could wipe out Edison's DC system. He already had a number of excellent engineers working for him, including a young man called William Stanley who understood transformers. Stanley designed the first AC lighting plant for Westinghouse and on 23 March

1886, it was switched on in Stanley's home town of Great Barrington, Massachusetts.

Westinghouse used his transformer patents to full effect. The current was generated at 500 volts and transmitted to the customers where a transformer lowered the voltage to a safe fifty volts before lighting lamps. Because the AC electricity system needed only a fraction of the copper wire that Edison's system needed to transmit the same power, it cost far less to set up and maintain, and the profit margin was much higher. The system would also give Westinghouse a tremendous advantage over Edison in remote locations.

Stanley had told him that an efficient transformer was the key to a practical AC power system, and perhaps it was, but Westinghouse knew that Edison provided everything a customer could possibly want with his DC system – lighting, heating and motors for mechanical power. The Westinghouse AC system had heaters, but no lights or motors. Just having a transformer would not be enough if Edison would not licence his incandescent light to his competitor.

Westinghouse had got to use Edison's incandescent light by very dubious means. He bought a company called United States Electric that held a set of incandescent lighting patents (which allowed him to make lamps that didn't work very well) from Sawyer and Mann. He then made copies of Edison's carbon filament lights claiming they were made under the Sawyer-Mann patents, even though the Sawyer-Mann design was greatly inferior to Edison's. Although Edison challenged Westinghouse in the courts, he did not succeed for many years in stopping Westinghouse making the lights. While the case was being heard, and thus not decided, Westinghouse could still claim the right to make the carbon filament lights that were identical to Edison's, but without paying any royalties to Edison. The importance of this long-running legal battle will become clearer as the story unfolds.

Westinghouse now proved himself to be as good at marketing as Edison. He turned his late entry to the market into a selling advantage. This is how he advertised his new AC lighting system to the people of Great Barrington:

> We regard it as fortunate that we have deferred entering the electrical field until the present moment. Having thus profited by the public experience of others, we enter ourselves for competition, hampered by a minimum of expense for experimental outlay. In short our organisation is free of the load which other electrical enterprises have. We propose to share the fruit of this with our customers.

However, if Westinghouse were to dominate the market he needed to supply industrialists who wanted motors, and, as mentioned earlier, he had no AC motors to offer. The invitation to Tesla's lecture came at just the right moment to serve Westinghouse's commercial self-interest. If Tesla really had made a practical AC motor, then Westinghouse's AC system would be able to oust Edison's DC; and he would be able to offer an electricity supply to industrialists that would be far superior to anything Edison had to sell.

Westinghouse came away from Tesla's lecture impressed – all Tesla's components: generators, transformers and motors had been carefully designed to work together. But he could see problems ahead: his Great Barrington plant used single-phase electricity with a reversal frequency of 133 times a second. This was much faster than the electricity that Tesla's motor used, which ran best when using two- or three-phase electricity reversing at between fifty to sixty times a second. But that problem, he consoled himself, could be sorted out – if he got Stanley and Tesla working together.

chapter 6

Old Sparky

On August 6, the New York Prison Authorities executed condemned murderer William Kemmler using the 'electric chair'. The electric charge was too weak and the miserable work had to be done again, becoming an awful spectacle, far worse than hanging.
New York Times, *August 7, 1890*

Just one month after Tesla's lecture to the Institute, Westinghouse made an appointment to meet him. They met in Tesla's laboratory just a few doors up from Edison's offices on Fifth Avenue. As he passed Edison's office, Westinghouse doubtless allowed himself a smile at the march he was about to steal on his rival.

Tesla, who was immediately impressed by Westinghouse, said afterwards: 'First impressions are those to which we cling. I like to think of George Westinghouse as he appeared when I saw him for the first time. The tremendous energy of the man had only in part taken kinetic form, but even to a superficial observer its latent force was manifest. Always smiling, affable and polite, he stood in marked contrast to the rough and ready men I met.'

Short, bearded and ten years older than Tesla, this polite but ruthless man appeared to Tesla to be strong and decisive, the type of authority figure he had always responded to in his youth. Instinctively he wanted to impress and please this man who had come to judge his invention. He demonstrated his equipment and Westinghouse immediately saw it as his chance to take the electricity supply industry away from Edison. He offered Tesla a million dollars for his AC patents.

Tesla considered the offer, then showed that he had learnt a little about business: 'One million dollars, plus a dollar per horse-power royalty, and you've got a deal,' he said.

He was learning, but not quickly enough. He had arranged a valuable royalty, but would he be able to collect it? He did not know that Westinghouse was making Edison lamps without paying royalties to Edison, and certainly didn't know that Westinghouse was pirating Edison's incandescent lamp design while hiding behind a deliberate misreading of the Sawyer-Mann patents.

Tesla had conceived his AC system as one big idea, but what he eventually sold Westinghouse were forty patents, working out at $25,000 each. This was about half the price Westinghouse had been prepared to pay.

Still, Tesla was sure that the royalties would earn him plenty of money.

Speaking many years later to the Institute of Immigrant Welfare, Tesla said of Westinghouse: 'George Westinghouse was, in my opinion, the only man on this globe who could take my alternating-current system under the circumstances then existing and win the battle against prejudice and money power. He was a pioneer of imposing stature, one of the world's true noblemen of whom America may well be proud and to whom humanity owes an immense debt of gratitude.'

To help get the machines into production, Westinghouse offered Tesla a job as a consultant at his Pittsburgh factory, and Tesla, although he would have preferred to stay in New York and continue his research, accepted. Now the fun could really start.

Once in Pittsburgh, Tesla had to work with William Stanley, an engineer who understood the commercial aspects of his work. Stanley knew that if Westinghouse was to succeed in his venture, Tesla's motor had to be turned from prototype into a saleable product quickly. If the on-going court battle with Edison was unsuccessful, then Westinghouse would be unable to manufacture electric lights, and the whole venture could fall apart. In 1888, Westinghouse had all the elements of a superb AC power system, but only if Stanley and Tesla could make them work.

Unfortunately the two engineers did not get on with each other. Tesla was used to working on his own and disliked working under pressure to deadlines. He had a vision of how his power system would be used for the service of the world, and, in particular, benefit people in remote locations such as Smiljan where he had spent his early childhood. He was driven by the conviction that if his mother had had access to the civilizing benefits of electric power she would have been a much more relaxed and loving person. Now he intended to make sure that other people got the chances his mother had not been given.

Knowing that the 133 cycle system that Stanley used was unsuitable for his motor, Tesla would not, as was usual when he knew he was right, compromise. Stanley, on the other hand, thought that the higher frequency current helped the transformers to work more efficiently and did not want to use Tesla's lower frequency. Tesla argued that his transformers worked at sixty cycles and were better than Stanley's. And this was not their only difference of opinion. Tesla wanted to use the two-phase current to make the motor design more elegant, but Stanley wanted to use the single phase to save money.

Eventually to keep the peace, Westinghouse suggested that Tesla might care to go back to New York and continue his research. Perhaps he could arrange some demonstrations of AC power to counteract the bad publicity that Edison was starting to cause. He could be hired as a consultant when and if he was needed. Tesla was glad to go back to his laboratory; Stanley was glad to see him go, thinking that he could now get on and complete the practical development of the Tesla patents.

It is clear from Tesla's writings that he did not enjoy his time in Pittsburgh. In his own words:

> *Great difficulties had still to be overcome. My system was based on the use of low frequency currents and the Westinghouse experts had adopted 133 cycles with the object of securing advantages in the transformation. They did not want to depart from their standard forms of apparatus and my efforts had to be concentrated upon adapting the motor to those conditions. Another necessity was to produce a motor capable of running efficiently at this frequency on two wires which was not easy of accomplishment. At the close of 1889 however, my services in Pittsburgh being no longer essential I returned to New York and began immediately to work on the design of high frequency machines.*

He later took great delight in the fact that Stanley was forced to use the lower frequency to make the system reliable.

Back in New York, Tesla now had a new interest. His arguments with Stanley had set him thinking about the effects of changing the frequency of electric current, and he was now wealthy enough to indulge himself and investigate this. He was enjoying playing the great inventor and planned to go on a tour of Europe. His mother was unwell and he wanted to share his success with her and hear her say that he had finally succeeded in matching the potential of his dead brother.

Never at ease in female company, least of all his

mother, Tesla did not know many women, and those he did know were mainly married to colleagues and associates. His dedication to work did not allow him much time for socializing and, anyway, he had trouble relating to non-engineers. Perhaps if he had met a woman scientist who shared his advanced views on electricity, he may have been attracted but he never met such a woman. However, before he could possibly spare the time to visit his mother he wanted to experiment with different frequencies of current.

Meanwhile, Edison had come up with a novel way of getting back at Westinghouse that would trap Tesla and his AC system as piggy-in-the-middle of the two ruthless businessmen. Using strong rhetoric, he had issued a public warning in his Bulletin describing deaths caused by the high voltages used in certain arc lighting systems, and had drawn attention to the low moral character of 'patent pirates' whom he accused of being hell bent on introducing dangerous currents into the homes of decent American citizens. He had become the electricity industry's Jeremiah predicting the destruction of Jerusalem unless people listened to him. Quoting him, the Bulletin reported: 'Just as certain as death, Westinghouse will kill a customer within six months after he puts in a system of any size. He has got a new thing and it will require a great deal of experimenting to get it working practically. It will never be free from danger.'

Such purple prose needs to be supported by evidence if it is to be believed, so Edison decided to back up his claims and demonstrate the death-dealing intentions of his opponents by carrying out high-voltage AC electricity experiments on animals in front of newspaper reporters and other invited guests. Charles Batchellor was delegated to shove a stray dog or cat on to a sheet of tin to which wires from an AC generator, supplying electricity at 1,000 volts, were attached, so that the watching press could be convinced of the lethal possibilities of being 'Westinghoused'. So many stray dogs and cats were purchased from eager schoolboys in West Orange, that it seemed that the local small animal population would soon become extinct.

Not all the model executions went smoothly. On one memorable occasion, Batchellor, the loyal executioner, almost died himself. In this test, the current had been connected to a bowl of water on a metal plate, but the puppy chosen to demonstrate the dangers of AC power refused to drink the lethal draft. In leaning forward to encourage it, Batchellor received an electric shock that threw him across the room. Badly shaken, he told afterwards of how his body seemed to be wrenched apart by a great rough file being pulled though it. He survived, but the puppy died a martyr to Edison's cause.

At this time, reformers in the field of capital punishment were pushing the State of New York to find and use a more humane method of execution than hanging. A Commission, consisting of Dr Carlos McDonald, Dr A.D. Rockwell and Dr Edward Tatum, and headed by Harold Brown, an 'electrical expert' who was a former laboratory assistant of Edison's, was appointed to investigate. Brown, who had assisted at some of Edison's lethal animal tests, realized the potential of AC electricity as a method of ensuring an 'instantaneous, painless and humane death'. And, as a result, Edison offered Brown the use of his laboratory to carry out experiments with 'Westinghouse current' so that he could advise the Commission from personal knowledge. Instead of simply repeating Edison's experiments, Brown also carried out a number of public executions of large dogs and an unfortunate horse.

In the autumn of 1888, the New York State Legislature adopted a law allowing the use of the 'electric chair' in place of hanging, as a means of capital punishment. The chair was soon named 'Old Sparky' by the popular press when Harold Brown made a very public purchase of three Westinghouse alternators as the most suitable equipment for electrocuting condemned criminals.

Meanwhile, Westinghouse desperately needed an equally public display of success for the Tesla system, and Lucien Lucius Nunn, a colourful country lawyer from the little mining town of Telluride, was about to give him that chance.

Nunn, a lawyer by trade, moved to Telluride, a boom mining town in Colorado in 1881, just as mining there was entering a fuel crisis. At this time, mines at high altitudes were using steam power for their operations, and those above the timber line, having used up their local timber, had to bring coal in by mule train. As the surface seams became exhausted, the miners had to dig deeper and needed even more power for pumping, winching and working the deeper seams. The cost of packing in coal by mule was pushing them towards bankruptcy.

One mine that was close to failure, although it still had good gold reserves, was the Gold King Mine, and in 1888 Nunn formed a consortium to rescue it. He knew that the problem was the cost of energy, but also knew that there was a limitless source of power in the San Miguel river. If used, the river could produce 2,000 horsepower, more than enough to meet the needs of all the mines in the area. The only snag was that the river was over 2½ miles (4 km) from the mine, too far to use Edison's DC system. An isolated Edison system was also of no use because that would have to be run by steam, and coal was costing $50 a ton to carry in. The only way to make the Gold King Mine profitable would be to tap into the free energy of the distant fast flowing river, but could this be done?

As luck would have it, Nunn's brother Paul was a member of the Institute of Electrical Engineers and had heard Tesla's lecture on AC systems and knew that Westinghouse was developing the Tesla patents. To his credit, Westinghouse already had a working AC power system in Massachusetts; and another member of the Gold King consortium Benjamin Butler, who had been previously Governor of Massachusetts, had seen for himself the success of Westinghouse's first AC system. In 1890, the Gold King Consortium went to Westinghouse and asked him to supply them with the first industrial Tesla Electricity Supply System.

This was a major test for the new Tesla machines that William Stanley was working so hard to adapt for mass production. The Gold King contract was for a six-foot (1.8 m) water-wheel driving a 100 horsepower generator that made electricity at 3,000 volts AC. A 2½ mile (4 km) power line was supplied to carry the current from the river, and up the hill to the mine where it drove a 100 horsepower motor. When Westinghouse took the contract, none of the electrical equipment had ever been made before. It was a total leap into the unknown, but if it worked it would be conclusive proof that Tesla's ideas were sound. Westinghouse publicized the contract hoping that this would counter the adverse publicity that Edison was creating for AC power. The contract was so important that Westinghouse was willing to spend $25,000 more than he was paid for it, to ensure success.

Edison, realizing that his bad publicity campaign was in danger of being short-circuited, encouraged a gruesome debate. Articles explaining how to use AC current to dispatch convicted felons started to appear in the press.

Harold Brown had been loud in his praise of Edison, complaining in the press that Edison's enemies had never forgiven the man for proving, by demonstration, that alternating current could produce death at very low voltage pressures. As the Commission Chairman, Brown had obviously taken a keen interest in his work and was able to go into great detail about electric-chair procedures in newspaper interviews. He knew how the executee, having had his head and leg shaved to allow the wires to be attached, would be strapped into the wooden chair with the metal cap clamped to his head and the metal plate strapped to his leg, all carefully wettened with potash solution to ensure good electrical contact. He described how when the switch was operated the felon would be beaten to instant death by the powerful contractions of his own muscles. 'Thus the majesty of the Law will be vindicated without physical pain,' Mr Brown added to reassure the squeamish.

However, it eventually became clear that Brown did not like AC current. 'Knowing the deadly nature of alternating current,' he later said, 'I cannot understand why the public did not demand that the legislature banish it from the streets and buildings, thus ending the terrible, needless slaughter of unoffending men.'

Westinghouse did not remain silent during this debate. He pointed out how sad he was that the struggle for control of the industry had become so bitter and personal. Despite his sadness, however, he then went on to make a personal attack on Edison, quoting him as saying, 'I don't care so much for a fortune as I do for getting ahead of the other fellow,' and 'my personal desire would be to prohibit entirely the use of alternating current.'

'This,' Westinghouse added, 'from a man who uses a dangerous 220 volts for his overhead public supply cables.' He then arranged a public demonstration where he roasted a side of beef in less than two minutes by passing an Edison DC current through it. Only the household pressure of 115 volts was used, he assured his witnesses.

He summed up by saying: 'Mr Edison has always said, that in the long run, every system will fail which does not for domestic service use low-pressure current. This is exactly what alternating current supplies.'

The public were completely bemused by this war of words, but they trusted Edison, and he was convinced that his best argument would be the public death of a convicted killer. Murderer William Kemmler was selected for the honour of being the first man to be dispatched by electric chair.

Westinghouse then paid the bills for Kemmler's lawyer to appeal against this new form of execution, and Edison was called into the witness box to give his opinion on the use of electric current to cause death. During cross-examination under oath, Edison was asked if he was testifying from belief or knowledge, and he replied belief, as he had never killed a man with electricity.

The appeal was lost and Kemmler's sentence of death by electrocution was confirmed. On 6 August 1890, it was carried out, but was not the swift pain-free death that Harold Brown had promised. The current applied was too small to cause instant death, so the hapless Kemmler spent fifteen minutes roasting slowly in a state of twitching cramp. Then, when the current was turned off, he was found to be still alive. So, it was turned on again, until, after an interminable period of smouldering and agonizing spasms, he was finally dead. Hanging would have been much quicker and kinder.

After this horror, which was reported in the *New York Times*, Westinghouse needed to restore public confidence in alternating current. To do so, he leaked a story to a reporter from *The Electrical World* magazine, of successful progress at Telluride. The reporter's scoop was published on 21 March 1891:

> *This Telluride plant promises to be . . . one of the most interesting mining plants in the world from the nature of its location, which made electrical power a necessity, and especially from the daring way in which the difficulty of using very high potential has been met by employing a synchronising motor.*

A few weeks later the Gold King Mine's new Tesla AC system was turned on and was an immediate success, running for thirty days and nights without interruption. This protracted trial period may, in part, have stemmed from the fear that if they turned it off it may not restart. But this fear proved to be groundless. The plant was reliable and safe. The public, however, would need more than a demonstration in an obscure little mining town to convince them of the virtues of AC current. Something more spectacular was called for.

Back in his New York laboratory, Tesla had become interested in the effect of high-frequency electric currents, and had discovered a subject called resonance, a natural effect that amplifies small movements and makes them larger. Imagine a child on a swing: the first push starts the child moving to and fro; as the swing changes direction, the child stops completely for a moment. If you continue to give the swing a small push each time it pauses before starting a new cycle, then the speed and movement will become faster and faster. Indeed, you will soon have to stop pushing to avoid the child being pushed right over the top. This increase in the amplitude of the swing by making small carefully timed pushes is called resonance.

Tesla realized that electrical circuits that have both coils and capacitors behave in a resonant way and that if he added just a small amount of electrical energy to this type of circuit, at just the right time, he could make very high voltages and very high-frequency currents. In a working note in 1890, he wrote:

> *The first question to answer then is whether pure resonance effects are producible. Theory and experiment show that such is impossible in nature for, as the oscillations become more vigorous, the losses in vibrating bodies and environing media rapidly increase, and necessarily check the vibrations, which would otherwise go on increasing forever. It is a fortunate circumstance that pure resonance is not producible for, if it were, there is no telling what dangers might lie in wait for the innocent experimenter.*

This thinking led him to invent the Tesla coil, a device that used resonance to create high-frequency, high-voltage electricity. At the same time, he also developed the coil-condenser tuning system that is the basis of all modern radio and television.

Tesla patented his 'Tesla Coil' and radio-tuning device six years before Marconi's first radio patent was issued, a matter I will return to later.

Tesla discovered a very important principle about high-frequency electric current – it does not travel through the middle of a wire or any other type of conductor; it travels through its outer edges. This effect, known to modern engineers as the 'skin effect', is made use of in steel wires with a copper outer coating; the steel makes the cable strong and the copper provides low resistance in the outer skin where the current travels. The 'skin effect' started Tesla thinking about the behaviour of waves, and he was the first scientist to realize that heat, light, radio, sound, alternating magnetism and alternating electricity, were all the same mathematically. If he could fathom one of these effects he could understand them all.

During this productive period of his life, Tesla also invented the gas fluorescent lamp which still lights most of today's offices and public buildings. They give very high light levels for small amounts of electricity and are much more efficient than filament lamps at turning electricity into light. He fitted out his laboratory with a great loop of wire around its outer wall and passed high-frequency current through this wire from a special alternator he had made. A series of gas tube lamps were placed around the lab to light up when he passed power through the loop. The lights could be positioned anywhere in the lab that he wanted to work, without the necessity of running a wire to them because they worked by 'wireless' power. He broadcast the power from his large loop and picked it up from a small loop of wire attached to the terminals of each of his gas discharge lamps.

Tesla carried out a number of dramatic experiments with his new high-frequency currents. He found that he could pass high voltages safely though his own body, provided he ensured that the current stayed low. He discovered what all electrical engineers who survive an electric shock know, that it is current that kills not voltage. Five thousandths of an ampere passed across the chest will stop the heart, but two million volts at a millionth of an ampere will just make your hair stand on end without harming you.

After his lecture and the sale of his patents to Westinghouse, Tesla became popular with the science writers of the New York newspapers. Westinghouse encouraged him to promote the safety of Tesla currents and Tesla was often invited to dine out by curious newspapermen. He never set up a household of his own, always living in hotels and eating in restaurants, so, to return hospitality, he invited groups of science correspondents to dinner parties and then took them back to his laboratory to show off his electrical 'magic'.

With his new found wealth, Tesla started to spend his evenings enjoying the social life of New York. Since living with his fastidious aunt in Carlstadt he had become a gourmet. Now that he had some spare money, he took to giving large dinner parties at fashionable venues, such as the Waldorf Astoria and Delmonico's restaurant. These parties grew in popularity and New York's social set clamoured for invitations. His liking for good food and his skills as a knowledgeable well-read conversationalist, plus his European manners and well publicized million dollar deal with Westinghouse, made his invitations desirable among the socialites. Some over-optimistic mothers even tried to interest him in their daughters. Tesla was flattered by the interest, but didn't strike up any romantic liaisons.

His well ingrained respect for authority figures and snobbish attitudes drove him to show off in front of these new-found 'friends', and his tendency to be pretentious and display his cleverness led him to give a series of bizarre demonstrations involving some awesome properties of high-voltage, high-frequency electricity.

He knew that if a voltage is vibrating at a very high frequency the current it carries would not pass through his body but go harmlessly along his outer surface. Using this knowledge and his own body as a conductor, he staged spectacular demonstrations of the safety of AC power. One of his most breathtaking tricks was to hold a wire from a Tesla Coil in one hand and draw a spark from the fingers of his other hand to light a lamp.

These scientific freak shows were given at his laboratory, as a climax to his dinner parties. He always dressed in a black tail coat and white shirt for these after-dinner shows, and sometimes even added a silk top hat to increase his already impressive height. He also sported platform boots which added a further 6 inches (15 cm) to his height and served as a rubber insulation under his soles.

Sparking and crackling round his darkened laboratory, making lamps glow from his pointing finger, he must have presented an awesome spectacle, must have looked like a modern god of lightning.

Pictures appeared in the social pages of the New York papers showing Tesla in morning dress, wreathed in sparks. A typical caption read:

> NIKOLA TESLA. *Showing the Inventor in the Effulgent Glory of Myriad Tongues of Electric Flame After He Has Saturated Himself with Electricity.*

The effects of Edison's claim about the deadly risks of AC electricity were being disproved in a most spectacular manner. Things were starting to look better for Tesla and his AC system, but the world of business still had another hard lesson in store for him.

chapter 7

Meet Me At The Fair

My mother was an inventor of the first order and would, I believe, have achieved great things had she not been so remote from modern life and its multi-fold opportunities.

Nikola Tesla

Men of business do not always cherish brainpower. 'Too clever for his own good,' they say. 'The motives and whims of scientists are just too impractical.'

Tesla's invention may have saved the mining industry of Telluride, but he hadn't been honoured by the town nor praised by the companies that had benefited from his ideas. Fortunately he was far to busy studying resonance in high-frequency currents to care about this lack of public acclaim. He hungered for praise, but didn't expect to get it, having learnt as a child that whatever he did would never satisfy his parents bereaved of their other son. This problem of low self-esteem was still with him even after the death of his father, but now his mother had become the focus for his imagined failures.

He was enjoying his research and the only worry that was niggling him was his mother's ill health. She was the only female he had any sort of relationship with, yet as the opening quote shows he seemed to value her more for her inventiveness than her motherly love. He decided that, as soon as he could spare the time, he would go to Europe to visit her. His success was making him feel guilty about not keeping in touch, but he didn't really want to face her. The frequent spankings that had rewarded his young experiments with pop-guns and swords carved from pieces of furniture still rankled as being 'not of the formal kind but the genuine article'. And it is clear from his 'autobiography' that, although he felt a duty to visit her, he was delaying this visit as long as he could. In his memoirs, he wrote:

> *A consuming desire to see her again gradually took possession of me. This feeling grew so strong that I resolved to drop all work and satisfy my longing, but I found it too hard to break away from the laboratory.*

His polyphase AC system had taken on a life of its own. Engineers throughout the world were beginning to see that his method was the only way to carry electricity over very long distances. As his reputation grew, Tesla continued with his new expensive experiments secure in the knowledge that, before long, Westinghouse would start paying him royalties. He was looking forward to a future of well-funded research and public acclaim as he built his better world for everyone, but things were not going quite as smoothly as he believed.

AC electricity was beginning to become respectable because it offered better profits for businessmen. Edison's DC system was losing its front place, but the most important lighting patents were still in Edison's hands who was remorseless in chasing companies which were stealing his patents, with Westinghouse the main object of his court-room battles. In 1891, the Edison company's long-running dispute with Westinghouse was coming to a head.

George Westinghouse's excuse for using Edison's lamp patents without paying royalties was that he had bought the Sawyer and Mann incandescent lamp patents. The problem was that their lamp never worked properly because it used a thick clumsy filament that burnt out in a matter of minutes, while Edison's lamp used a fine filament that lasted hundreds of hours. So, Westinghouse made Edison lamps and pretended that his crude Sawyer-Mann patents gave him the right to do so. Edison's Lighting Company had taken the matter to court, but the legal argument had been dragging on for over six years and it looked as if it might not be settled before the patent expired in 1894.

Westinghouse was not the only company trying to rip off Edison's patents. The Thomson-Houston company, run by an ex-shoe salesman, was also breaking the Edison patent and was co-defendant in the long-running court case. The incandescent lamp patent, along with all the other Edison electrical patents, was held by the Edison Electric Light (EEL) Company. This company was now controlled by Edison's bankers, and two of the most important bankers on the EEL Board were JP Morgan and Edward Adams. As bankers, these businessmen were more concerned with profit than with technical issues. They had not supported Edison when he set up factories to make electrical fittings, but had got involved after he had made a success of manufacturing.

Edison was suffering the problem that many growing businesses have, he was running out of money. Because his business was successful he was getting more orders, but he wasn't being paid for the orders until he delivered the equipment. Meanwhile, he was having to find the wages to pay 2,000 people every week, and was buying his raw materials long before he got paid for the finished goods. The more work he got, the more money he had to borrow to stay afloat.

By 1889, he was in serious debt and having trouble paying his employees' wage bill. If Westinghouse and Thomson-Houston hadn't been taking business off him, using his patents, then perhaps Edison wouldn't have been so vulnerable. Whatever the cause of his financial difficulties, the bankers saw a chance to take over Edison's manufacturing business.

The Edison Electric Light Company had been just a paper company for collecting royalties on Edison's patents, but now the bankers who ran it offered to buy Edison's factories. Negotiations went on for a long time, but finally Edison agreed to sell out for $1,750,000. He seemed relieved that somebody else was taking responsibility for the weekly wage bill and took his first holiday in years, a trip to Europe. The bankers' new firm was called the Edison General Electric Company. This merger of a large part of the electrical industry made smaller companies vulnerable when other bankers started to see the advantage of merging smaller companies into bigger units, whether the owners wanted to or not.

One of the banking directors of EEL was worried about the long-term viability of the DC power system. This man, Edward Adams, also happened to be a friend of George Westinghouse. Realizing that between the two of them, Edison and Westinghouse held every important patent for the future of electricity, he tried to get the two of them together to settle their differences, but Edison sent him a forceful reply:

> *I am very well aware of Westinghouse's resources and plant, and his methods of doing business are lately such that the man has gone crazy over sudden accession of wealth or something unknown to me and is flying a kite that will land him in the mud sooner or later.*

Twenty years later, however, Edison admitted that he had been wrong to turn down Adams' suggestion of co-operation with Westinghouse. If he had not, the costly and abusive 'Battle of the Currents' might have been avoided.

In 1889, the legal action came to a head in the court room of Mr Justice Bradley of the United States Circuit Court of Pittsburgh. Arguments had revolved around a counter-claim that Edison had not been the first to invent the incandescent electric light, and that his 1879 patent did not give sufficient information for a mechanic to make such a lamp.

Justice Bradley saw that earlier inventors had made incandescent lamps, but that those lamps were failures because they only stayed alight for a few minutes and, because of this, had no commercial value. His legal verdict read as follows:

> It seems that they were following a wrong principle, the principle of small resistance in an incandescing conductor, and a strong current of electricity; and that the great discovery in the art was that of adopting a high resistance in the conductor with a small illuminating surface and a corresponding diminution in the strength of the current. This was accomplished by Edison.

Bradley used Ohm's law on Edison's behalf to support the US Patent Law. The idea of making a thin filament which only needed a small current to make it glow was the true step forward. Edison had used Ohm's law to advantage and this was the basis of his patent, so on 4 October 1889, Justice Bradley ruled in favour of Edison.

This decision put both Westinghouse and Thomson-Houston in a difficult position. By now, Edison had sold out his interest in the new Edison General Electric Company, although he still managed it, and the bankers running the company were going to drive a hard bargain. Soon it would be a case of either paying royalties or closing down their lamp-making factories. To play for time, Westinghouse and Thomson-Houston appealed to the Federal Courts.

Edison was frustrated by the delaying tactics of the 'pirates' who continued to thwart his business, and his unscientific attacks on the deadly properties of his opponent's AC systems now took on extra venom.

Ohm's law, too, was proving a fickle mistress. It had given him a chance to win his legal battle, but it was also damning any long-term future for the power systems on which he had built his company. Soon it would play a further trick on him. Justice Bradley had used it to support Edison's patent but DC power still would not travel as far as AC.

Westinghouse's appeal was heard in New York. He used a team of scientific experts and top lawyers to publicly humiliate Edison. Westinghouse reasoned that if the success of Edison's patent relied on the application of Ohm's law, then he would show in open court that Edison did not understand that law. It must have gone through his mind that if he could show that Edison did not really understand electricity, the great man's view on the dangers of AC power would carry less weight with the public. He could then use Edison's testimony, along with the news of the successful Telluride power station, to promote AC power. Small wonder that he had been prepared to build Telluride at a loss. Its triumph clearly showed that Edison did not understand electricity when he spoke out against high-voltage pressure for long-distance transmission.

The Telluride plant had been running continuously for a month when Edison stood in the witness stand before Judge William A. Wallace. The cross-examination on that fine June morning was intended to expose Edison's ignorance of the theory of electricity. The record of the trial shows in excruciating detail how Edison was humiliated.

> *'What did you know of Ohm's law when you set out to make your filament lamp, Mr Edison?' asked the lawyer.*
>
> *'I did not fully understand everything about Ohm's law when I started work in 1878, for had I done so it would have prevented me from experimenting,' the inventor replied.*
>
> *'Why should a knowledge of Ohm's law prevent you from experimenting?'*
>
> *'Because I would try to figure it out mathematically, and I have had a great many mathematicians employed by me for the last ten years, and they have all been dead failures.'*
>
> *'But was it not the laws of electrical science which showed you the way to the high resistance burner?'*

'I don't think so. The mathematics always seems to come after the experiments, not before.'

'Mr Edison, did you make use of mathematicians to assist your work?'

Edison replied. 'I can hire mathematicians, but they can't hire me.'

Having shown Edison's lack of theoretical knowledge, the lawyers put his hired mathematician on the witness stand who, despite intense cross-examination, totally defended Edison's experimental results. Westinghouse had succeeded in humiliating Edison, but had not shaken his case. There was only one faint hope left. His lawyer tried to claim that Edison's original patent specification was impossible to follow to make a working incandescent lamp, so the patent must be invalid.

Ever the practical man, Edison sent for a technician and had him make a lamp by following the instructions while the court officers watched. The lamp was still burning after 600 hours. Edison's case was proven and Judge Wallace ruled in his favour.

The case had cost Edison General Electric over two million dollars and left Westinghouse almost completely broke, yet the 'Battle of the Currents' was not entirely settled. The court decision had left Westinghouse without an incandescent lamp to sell in any future power systems, and it seemed that he would have to wait two years until the Edison patent expired before he could again make incandescent lamps. Given the state of his finances, that might be too late.

Although Edison General Electric was also in debt, it still had control of the lighting industry for at least two years. Charles Coffin, the ex-shoe salesman who ran Thomson-Houston (TH), saw only one way out. He went to JP Morgan, who controlled Edison General Electric (EGE) and proposed that merging the two companies would solve two problems in one move. His reason was that, due to the ruling, Thomson-Houston could not make lamps, but Edison General Electric was also in difficulty because it was in debt to the tune of $3,500,000. Despite its legal battle, TH had very strong finances. Putting the two companies together would give them three quarters of the US lighting market and a strong funding base. The deal was done. Coffin became the new chief executive, and Edison's name was dropped from the new General Electric (GE) company.

The headlines in the New York Papers read:

MR EDISON FROZEN OUT HE WAS NOT PRACTI-CAL ENOUGH FOR THE WAYS OF WALL STREET.

The merger may have removed Edison from the scene, but Westinghouse was also now in serious money trouble. GE now had an enormous chunk of the existing lighting market, but Westinghouse had the Tesla patents that were the key to the future. He only had a few cards in his hand, but they were strong ones if played well.

He went to his bankers who suggested a solution which meant, like Edison before him, Westinghouse was about to lose his company to businessmen. Both inventors had ignored the basic rule of business: growing companies need money and rapidly growing companies need lots of money. Having ideas and selling to people is only part of the secret of success, sound financial control is also needed. Westinghouse's bankers were about to teach him this lesson the hard way.

They offered to bankroll the formation of a new company that would involve many of the remaining independent firms. The US Electric Company and the Consolidated Electric Light Company would merge with the Westinghouse Electric Company to form the Westinghouse Electric and Manufacturing Company. But there was a catch in the deal. The Tesla patents were the main asset that Westinghouse had to offer, but his bankers didn't want the ongoing royalty deal with Tesla because it was too expensive.

Westinghouse got Tesla to sell his royalties outright for $216,000. Some romantic writers have hinted that Tesla gave them up in a grand gesture to ensure the success of his invention, but the truth is less sentimental. Tesla wrote to a friend that he had believed that the waiver of the royalties was only a temporary arrange- ment and that he needed the lump sum for experiments. Indeed, when he gave evidence under oath in a court case some years later, he said he was ignorant of the details of the deal as he always left such matters to his business partners. Doubtless this is why he never managed to keep what money he made and spent a lot of his time in debt.

Shortage of money was a recurring problem for Tesla.

The experiments he wanted to carry out were going to cost more money than he had left; and getting on with his tests was his only interest. When Westinghouse approached him to buy his royalties, he could probably see no further than the next experiment. The $216,000 Westinghouse offered him was more than enough to pay for this and for his trip to Europe as well. The so-called 'temporary arrangement' misunderstanding cost him $12,000,000, but saved Westinghouse's business. Maybe Westinghouse didn't try too hard to clarify the options for Tesla; or maybe Tesla, feeling guilty about his mother, simply wanted the money to go to visit her. Whatever the truth, Tesla neglected yet another sound business principle – always read the small print before you sign, even if distracted.

The Telluride project had shown that AC power could work reliably, but a small industrial project in a little mining town was not enough to overcome the bad publicity of 'Old Sparky'. If AC power was to be shown to be safe, then a much bigger gesture was needed. As it happened, a World Fair was to be held in Chicago, and its organizers, who were looking for a lighting system, were calling for bids to supply about 90,000 incandescent lights to light the show ground.

General Electric was quite sure that it would get the contract because it had court orders stopping Westinghouse E&M from infringing its patents when manufacturing lights. So sure was the company, that it put in a bid price of $18.50 per lamp using DC power. At this point, Westinghouse decided to take a gamble: losing the patent case might have meant that he was no longer allowed to make carbon filament lamps in one-piece glass bulbs, but he reckoned that if he made lamps with metal filaments in stoppered bottles he wouldn't be breaking the GE patent. Offering a Tesla AC system, he put in a bid for the job at $4.32 per lamp, less than cost but worth, he reasoned, the publicity; and, in doing so, offered to supply a product he hadn't even got.

He won the contract, but had to give guarantees to meet any losses suffered by the Fair organizers if he failed to deliver. If he failed, he would be ruined; even the extra income he had got from Tesla's royalties would not be enough to save him. With less than a year to achieve what he had promised, he put all his efforts into making a working two-piece stopper lamp with an iron filament.

While the 'battle of the currents' raged all around him, Tesla, undisturbed by its commercial implications, was busy satisfying his curiosity and looking at the effects of high-frequency current. The study of resonance was fascinating him, and he kept remembering an incident from his youth when he had been climbing a steep snow-covered mountain with some friends. To amuse themselves, they were playing a game where each, in turn, tried to roll a snowball further than the others. Suddenly one of the balls seemed to pass a natural limit. Instead of stopping, it rolled faster and faster, gathering more and more snow, until it became huge and plunged thunderingly down into the valley causing a tremendous avalanche. On that day, Tesla had witnessed resonance in action, and was now convinced that this could be applied to electricity. To prove it, he made all sorts of devices to produce different frequency currents for his tests.

He also thought that electricity could be used for medical purposes and, to try out these ideas, decided to experiment on himself. In a way that was certainly foolhardy, he passed high-frequency currents through different parts of his own body so that he could observe the effects. When he passed currents through his head, he noticed a pleasant feeling of drowsiness so he wrote an article suggesting that high-frequency current might one day be used as an anaesthetic. He was right about the effects of electric fields on the brain, but didn't realize quite how dramatic the effects of this type of equipment would prove to be. Modern medicine, with its increased knowledge of the electrical workings of the human brain, now uses this type of treatment – known as electro-convulsive therapy – to treat mental illness. But one of the possible side-effects of this therapy is loss of memory – something that Tesla was to discover, quite by accident, in very distressing circumstances.

Soon after selling his royalties to Westinghouse, Tesla decided to go to Europe. Accepting an offer to speak to the British Institution of Electrical Engineers, in London, he laboured hard over the lecture, which was a tremendous success. He next travelled to Paris to give a lecture to the French Electrical Engineers, which once again was a great success. Finally, he set off to visit his sick mother.

When he arrived at his mother's bedside, she was dying. It is unclear from medical records what her illness was, but the descriptions of her pain and suffering during the last six weeks of her life suggest cancer. Tesla remained by her side throughout those last six weeks, only going back to his hotel to sleep.

His mother had come from a long line of inventors; and both Tesla's maternal grandfather and great grandfather had invented 'numerous implements for household, agricultural and other uses'. Sitting at her side, Tesla remembered how she had caught him staying up at night to read his father's books by the light of a candle that he had secretly made for himself. She had spanked him, but had not reported him to his father. He remembered how when she discovered that he had lost all his money gambling – and yet was still craving another game of cards – she had given him a wad of notes, saying, 'Go and enjoy yourself. The sooner you lose all we possess, the better it will be. I know that you will get over it.' She had been right. She made him feel so guilty he had never gambled again.

He remembered walking with her in the mountains as a storm approached. The sky was heavy with clouds, but rain had not started when, all of a sudden, there was a flash of lightning followed by a sudden deluge. He had spoken to her about how the two phenomena were closely related, the lightning causing the rain to start. They had discussed the stupendous possibility of controlling the weather and how this would be the most efficient way to harness the sun to man's use.

Now, as he sat by the bedside of the only woman who had ever inspired him, he could only watch helplessly as each rattling breath became a greater pain for her. The spankings were forgiven as he realized she was about to leave him for ever.

Believing his mother to be a woman of genius, with great powers of intuition, he was convinced she would not die without trying to warn him, and, exhausted from his bedside vigil, gave in when his sisters insisted he should go back to his hotel to sleep. Afterwards, he said of that night:

> During the whole night every fibre in my brain was strained in expectancy, but nothing happened until early in the morning, when I fell into a sleep, or perhaps a swoon, and saw a cloud carrying angelic figures of marvellous beauty, one of whom gazed upon me lovingly and gradually assumed the features of my mother. The appearance slowly floated across the room and vanished, and I was awakened by an indescribably sweet song of many voices. In that instant a certitude, which no words can express, came upon me that my mother had just died.

As he awoke, he worried about this strange vision but recalled that he had once seen a painting that had allegorically represented one of the seasons in the form of a cloud, with a group of angels who seemed to be floating in the air. That same image had appeared in his dream, but with his mother in centre of it. The music had come from the choir in the church nearby at the early mass of Easter morning. He was satisfied that he could explain the entire vision scientifically, but as he got out of bed a messenger arrived to say that his mother had just died.

The strain of being unable to do anything except watch her die, must have been enormous. Speaking at her funeral, he said: 'She was a truly great woman, of rare skill, courage and fortitude, who had braved the storms of life and passed through many a trying experience. I must trace to my mother's influence whatever inventiveness I possess.'

After her funeral he collapsed and lost his memory. He had been testing his own electrical anaesthetic on himself and may have been using it as a sleep-inducing aid during times of stress. Because of his self-experiments, overwork writing lectures and the death of his mother, he had a breakdown.

As he recovered his health, his memory slowly returned, but he had become even more introspective. Friends said of him that he never quite left his world of engineering problems, often drawing sketches on the tablecloth while waiting for a meal or interrupting conversations to speak of experiments. Later, in an article entitled 'My Inventions', he wrote:

Upon regaining my health, I began to formulate plans for the resumption of work in America . . . I felt that

I should concentrate on some big idea . . . The gift of mental power comes from God, Divine Being, and if we concentrate our minds on that truth, we become in tune with this great power. My Mother had taught me to seek all truth in the Bible; therefore I devoted the next few months to the study of this work. If we could produce electric effects of the required quality, this whole planet and the conditions of existence on it could be transformed . . . The consummation depended on our ability to develop electric forces of the order of those in nature.

It seemed a hopeless undertaking, but I made up my mind to try it . . . after a short visit to friends in Watford, England; work was begun which was to me all the more attractive, because a means of the same kind was necessary for the successful transmission of energy without wires. At this time I made a further careful study of the Bible, and discovered the key in Revelation.

As a self-imposed penance for failing to be with his mother when she died, Tesla set himself the task of reading the whole Bible, and took two months to do this, reading full time. He never made clear to others what he actually found in the Book of Revelation, but, from then on, became fascinated with the workings of natural forces – something that would eventually lead him to harness the power of the Niagara Falls and much more.

He returned to America just in time to help Westinghouse prepare for the World Fair. The Pittsburgh factories had succeeded in making a lamp that did not infringe the GE patents. Everything was coming together. Westinghouse's gamble had come off.

When the Fair opened on 1 May 1893, there were 96,620 Westinghouse incandescent lamps, powered by Tesla generators, lighting up the fairground. At the Fair, Tesla exhibited and demonstrated a metal egg on a velvet platform. When he turned on the electric current, the egg stood on end and rotated rapidly, powered by the magic of AC. The crowds flocked to see him in his top hat and tails and high rubber boots, and watched him pass millions of volts of high-frequency electricity through his own body and light lamps with the sparks from his finger tips. Edison's lie about the inherent danger of AC power was finally dispelled before the masses pouring into the Fair. The 'battle of the currents' was reaching its final stage.

chapter 8

Harnessing Niagara Falls

Since electrical transmission of energy is a process much more economical than any other we know of, it necessarily must play an important part in the future, no matter how the primary energy is derived from the sun. Of all ways the utilisation of a waterfall seems to be the simplest and least wasteful . . . (it) implies no consumption of any material whatever.

Nikola Tesla

Part of the United States's northern boundary with Canada runs along the Niagara river. This river connects Lake Erie to Lake Ontario and sets the natural limits of the states of New York and Ontario. It starts in Lake Erie, by the town of Buffalo, and flows just over 33 miles (53 km) until it drains into Lake Ontario. The water level of Lake Ontario is 326 feet (100 m) lower than Lake Erie, and Lake Ontario is at the bottom of a great limestone cliff that the Niagara river gushes over. The river falls a spectacular 180 feet (55 m) at the spot known as Niagara Falls where the force of the falling water makes a perpetual mist of rainbows as the sun shines though the scattered water droplets thrown into the air.

The Falls are an awesome sight. As the Niagara river splits around Goat Island it tumbles with a breathtaking display of power over the cliffs, almost 900,000,000 cubic feet of cascading water power every hour. It's a famous tourist attraction for both the USA and Canada and a rich source of energy. For over 200 years, ever since Dan Joncaire built the first sawmill, driven by a water wheel turned by the Niagara's flow, the river has been a source of power.

Tesla knew about the force of running water. As a strong swimmer in his youth, his feats had sometimes outrun his ability, and he had almost died at the age of sixteen when he swam alone in a local river, upstream from a high mill dam in Carlstadt. He had failed to notice that the river was in spate, and that, instead of the normal 2-3 inches (5-7 cm) of placid water above the dam, there was over 12 inches (30 cm) of fast-flowing foam. As he swam out towards the middle of the reservoir, the current caught him and swept him towards the dam. It was over 50 feet (15 m) high, and if he had been carried over the drop it would have meant certain death.

He tried to swim against the current, but was swept on to the dam wall where he had the presence of mind to grip hold of it. Having done so, he was pinned like a butterfly in a glass case. The force of the water was crushing him, but if he let go the deadly pressure would carry him over the drop and smash him on the rocks below. He hung on with grim determination, desperately trying to think of a way out. It was no good shouting for help, there was no one to hear. He couldn't swim against the water flow and its force was crushing the breath out of him. To loosen his grip, in the midst of such a torrent, would end his life.

Afterwards, Tesla said that the danger of the situation forced him to think clearly. If the water was pressing him against the wall, he reasoned that he had to make himself smaller so that there would be less of him to crush. How could he do this? His answer was simple but inspired: turn sideways to the tide. This cut the pressure down by a third and, by keeping sideways to the flow, he was able to pull himself to the side and arrive exhausted on the bank. He had just been given a lesson he would never forget of how strong moving water can be.

After this, he was always fascinated by the power of water, and when he found a picture of the enormous Niagara Falls recognized in them limitless power. He wrote this school-day memory in the magazine *Electrical Experimenter* in 1919:

> *In the school room I turned my attention to water turbines. I constructed many of these and found great pleasure in operating them . . . I was fascinated by a description of Niagara Falls I had perused, and pictured in my imagination a big wheel run by the falls. I told*

my uncle that I would go to America and carry out this scheme . . . Thirty years later my ideas were carried out at Niagara.

The town of Niagara knew that it was sitting close to an enormous source of free power, but didn't know how to make use of it. In 1886, the citizens set up a seventeen-man Commission under the chairmanship of Lord Kelvin, one of the most famous scientists of the age, and, at the time, the world's most highly regarded independent authority on electricity. A Professor at the University of Glasgow, Scotland, Lord Kelvin was rightly renowned for his discoveries in electricity and magnetism. The job of his Commission was to investigate and report on the possibility of using the great waterfall at Niagara to make electricity. The Commission offered a reward of $20,000 for practical plans that could be implemented by a specially set-up company that was provisionally called the Cataract Construction Company. By 1891, the Commission had received seventeen plans for creating electric power from the Niagara river, but had rejected them all.

Westinghouse was too much of a businessman to supply plans simply for a prize. He knew that General Electric had not submitted a plan for the Falls because it could not use the Tesla patents; and that only one factory, in Niagara itself, had promised to buy electricity. The Pittsburgh Reduction Company wanted to use the power to smelt aluminium in electrically-heated furnaces. If, however, the system was to be worth building, it had to be able to transmit electricity to the industrial town of Buffalo, 22 miles (35 km) from the Falls. As Ohm's law prevented DC current being transmitted over that distance, Westinghouse – realizing that he would be stretching his finances and engineering resources to the limit if he tried to go it alone, and urged on by the advice of his Board to work with General Electric if he wanted his company to survive – decided to do a deal with General Electric and license the Tesla patents to them. This would make it possible for the two companies to submit a joint bid for a power station at the Falls. There were two major parts to the proposal; three 5,000 horsepower water-driven generators, the water piping and power house; and the transmission line to carry the power to a distribution system in the city of Buffalo.

Under this scheme, Westinghouse could earn income from royalties, Tesla's royalties, which General Electric would pay to the Westinghouse Electric Company. This stroke of business genius tied up the US electricity industry for the next 100 years, and Westinghouse Electric's lawyers enforced the Tesla patents so tightly that nobody else could get a toehold without paying for it. It made Tesla's name a byword for legal action, even though, having sold his enormously profitable royalties to Westinghouse for a trivial sum, he had no financial interest in the patents. The outcome of the game that Westinghouse played with his only asset, the Tesla AC patents, eventually made Westinghouse Electric and General Electric the two biggest companies in the world.

Edison's company had left him behind as it embraced the new Tesla technology of AC electricity, but that move had ensured both the company's future and the success of the AC system. The 'battle of the currents' was finally over and the two opposing companies were both winners. The only losers were Tesla and Edison – the two inventors who could not live with each other or cope with the ways of businessmen. Edison, realizing what had happened, went off to develop new interests in mining, but Tesla didn't realize for some time how shabbily the business world was treating him.

General Electric went from strength to strength. *The London Sunday Times* reported on its continuing success in November 1997, under the headline:

GENERAL ELECTRIC POWERS INTO EUROPE ON ALL FRONTS

As head of General Electric's lighting business in Europe Mike Zafirovski is following in the footsteps of Thomas Edison, inventor of the electric light and founder of the world's biggest company by market worth. But Zafirovski also has a leading role in a new revolution. For most of the 105 years since Edison started GE, it has been seen as an American leviathan with a growing number of overseas interests. Today, that image is undergoing a sea change. GE . . . is creating an elite corporate cadre that will dominate world business in the 21st century.

It certainly seems that the businessmen made a better

job of running the electricity companies than the inventors, and the companies that Edison and Westinghouse started still retain their original names and are alive and respected today.

In typical gambler's fashion, Westinghouse had told the people of Niagara that their electricity supply would be a sure-fire success. Fortunately, his first co-operation with General Electric did work out well. He chose a Scots engineer, George Forbes, to design and build the power station. The cautious Forbes was amazed at the 'typically American boldness' of the capitalists and manufacturers. Inspired by Westinghouse's 'hype', they invested large sums of money building factories to use hydro-electricity, long before the Niagara Falls power station was anywhere near ready.

Forbes designed a power station that used three 5,000 horse-power Tesla generators. He was unusually environmentally aware for those times because, in his design, he tried 'not to injure the natural beauties of the spot'. He built a canal from upstream of the Falls to carry the water to the power house where it was then taken through a 7 ½ feet (2.2 m) diameter pipe into the turbines. After spinning the great alternators, the water then travelled along a tunnel under the town of Niagara to discharge back into the river below the Falls.

To house workers for the new factories, a new industrial town was built alongside the Falls. It had a modern sewage works, electrical pumps to supply clean water, electrically-lit streets and well-paved roads. Built as a model 'smokeless' manufacturing town, it was completed at the same time as the power station. The prosperity of the region was then secured when, on 20 April 1895, the Niagara power station started to make hydro-electricity from the power of the water.

The newspapers described the project as 'the greatest engineering works in the world'. With the 'battle of the currents' over, Tesla's AC system had won the day. The final victory came the following year when General Electric, the company that Edison had formed, completed the power line to Buffalo and paid Westinghouse for the privilege of using the patents that Edison had rejected when Tesla had tried to give them to him ten years earlier. Edison's $50,000 joke at Tesla's expense had eventually cost him both his company and his reputation. At least Tesla had the last laugh.

The press swarmed around Tesla, praising and describing him as 'Nikola Tesla Our Foremost Electrician – Greater Even Than Edison'. Edison's harsher press coverage is summed up in the following comment: 'Edison was a bold and courageous innovator. Now he has become a cautious and conservative defender of the *status quo*.'

In reality, Edison had acted towards Tesla in the same way that the gaslight companies had acted towards him, and probably for the same reason: the fear that all the effort, equipment and capital invested in DC would be lost if AC replaced it. Unfortunately, the essence of great scientific work is that it is transitory and will always be replaced by new discoveries. Edison had been the great hero, but he was now replaced by the new idol, Nikola Tesla.

Such adulation can be addictive, and Tesla was just about to get a taste of a fame that would eventually turn him into a bitter disliked self-publicist. The vulnerable little boy residing within him had not expected praise, the successful engineer experiencing it for the first time could not get enough. If he had been less concerned with how others viewed him, perhaps he would have achieved the public approval he craved. Instead, although remembered by engineers and scientists, the mass of people who benefited, and continue to benefit from his genius, do not know his name. How this happened is a sorry reflection of his own failings both in business and personal relationships.

In the event, his fame could not have come at a worst time for someone with a hunger to be accepted by 'fashionable folk'. Unluckily for him, his first major success came during a period when newspapers were indulging in ever more lurid copy to attract readers. As a result, Tesla's reputation, in later life, suffered greatly from his association with outrageous schemes that never happened and from being unable to resist a grandiose quotation if it got him a headline. But this is a later part of his story. For the moment, we can enjoy the fruits of his victory in the 'battle of the currents', even though these were sowing the seeds for his later follies.

The Niagara Falls Power and Conduit Company, as the project was finally named, decided to celebrate the introduction of hydro-electric power to the city of Buffalo by holding a banquet at the Ellicott Club, the 'poshest' place in town. They invited distinguished guests from around the world to celebrate this triumph of American capitalism. After the banquet, the main toast of the evening was to be to Electricity, and Nikola Tesla was invited to respond.

Unfortunately, he treated the occasion as if it were a lecture to a learned society, and set out to describe a vision of the future, outlining the development of wealth in cities, the success of nations, and the progress of the whole human race, all powered by electricity.

The meal was sumptuous and the wine flowed freely that evening of 12 January 1897. It was getting late when Tesla rose to speak in the banqueting hall in Buffalo. He spoke of man's need for energy and how this need had been fulfilled by the steam engine until the electric motor had replaced it. He spoke of the importance of the electric motor for the future of industry, and the need to be able to generate electricity cheaply and efficiently. He told them his own thoughts on turbine design, and of the rapid advances in new applications of electricity, such as X-rays, welding, electric railways, the telephone and lighting. He praised the scientists who had made these advances possible: 'Nor is the work of these gifted men nearly finished at this hour,' he added. 'Much more is still to come.'

There was also much more of his speech to come. He described Niagara as a monument of enlightenment and peace, saying that it showed the subjugation of natural forces to the service of man: 'No matter what we attempt to do, no matter to what fields we turn our efforts, we are dependent on power. If we want to give to every deserving individual what is needed for a safe existence of an intelligent being, we want to provide more machinery, more power.'

His audience fiddled with their empty wine glasses, wondering when the port would be coming round. Undeterred, Tesla went on to talk of new ways of making electricity, of how the natural resources of the earth should be tapped for the benefit of mankind. He told of how he was coming close to solving the problem of making free electricity using the electrical charge of the earth itself.

He went on to talk about the need to transmit electricity throughout the world. As his lecture continued, the more restless members of his distinguished captive audience may have missed this throwaway comment: 'I have devised means which will allow us the use of power transmission of electro-motive forces much higher than those practicable with ordinary apparatus. In fact, progress in this field has given me fresh hope that I shall see the fulfilment of one of my fondest dreams; namely, the transmission of power from station to station without the employment of any connecting wire.'

The banker JP Morgan was in the audience and must have wondered how Tesla expected to make a living from free electricity broadcast to the world, but he couldn't ask because Tesla was still talking. Certainly, Tesla had just hinted at a very worrying scenario for Morgan to think about. The only way to deliver electric power at that time to a user was to run a wire from the power station to the customer. It was then clear who the supplier should bill for the power because the customer had strands of copper wire running into his building. If they didn't pay, they could be disconnected. But Tesla had just suggested that electric power could be thrown out into the air, so that anybody could collect and use it. This would certainly save money on the cost of the copper used to make the connecting wires, but the idea was sounding alarm bells in the business mind of JP Morgan. Was Tesla really saying that anybody anywhere could just put up an aerial and run their home, or even their factory, from electric power they collected out of the air? How would the power companies earn their money? And if the users didn't pay, how could they be disconnected without turning off everybody's power? Morgan, suddenly realizing that Tesla was still speaking, put these thoughts aside to listen.

Tesla was approaching the end of his forty-five minute marathon and intended to finish on a note of praise for the city that was hosting this splendid chance for him to show off. 'It is a pleasure to learn of the friendly attitude of the citizens of Buffalo and of the encouragement given to the enterprise by the Canadian authorities. We shall hope that other cities will soon follow Buffalo's lead. This fortunate city herself is to be congratulated. With resources now unequalled, with commercial facilities and advantages such as few cities in the world possess, and with the enthusiasm and progressive spirit of its citizens, it is sure to be become one of the greatest industrial centres of the globe.'

He sat down to rapturous applause. He had spelt out

his vision of the future, a vision that turned out to be surprisingly accurate in some areas and wildly optimistic in others, but he had planted a seed of doubt in the mind of JP Morgan who controlled General Electric. Edison's old company had just been forced to buy the right to use Tesla's AC patents to continue to sell electricity. Was Tesla now trying to invent a system that would give away free electricity and put General Electric out of business?

chapter 9

Resonance and Radio

The problem of producing light has been likened to that of maintaining a certain high pitched note by means of a bell. It should be said a barely audible note; and even those words would not express it, so wonderful is the sensitiveness of the eye. We may deliver powerful blows at long intervals, waste a good deal of energy, and still not get what we want; or we may keep up the note by delivering frequent gentle taps and get nearer to the object sought by the expenditure of much less energy.

Nikola Tesla

Half a million dollars is a lot of money for anybody, and for Nikola Tesla, when George Westinghouse bought his AC patents for this sum in 1888, it had been the price of freedom. He had to share some of the money with his partners, but at least half was his to do with as he liked.

Totally single-minded in his determination to study and invent, he was now looking forward to getting on with his experiments. But, first, before he started to enjoy his new wealth, he had to honour the pledge that was part of the deal, to spend time in Westinghouse's factory 'helping' to develop his machines.

A year later, both he and the Westinghouse engineers were glad that sufficient progress had been made to allow him to leave Pittsburgh and return to New York. Talking and arguing with the Westinghouse engineers had set him thinking about the speed of reversal of AC electricity, and he had a new idea that he wanted to play with. Why did the rate of reversal matter so much? The speed that the current changes direction is called its frequency. So far, his work had involved frequencies of no more than sixty reversals (or cycles) a second, but ever since his argument with Stanley about what frequency of AC should be used to power his motors, Tesla had been asking himself why this mattered? He knew that motors ran better at lower frequencies and that transformers preferred higher ones. Now, he decided to investigate the reasons for this difference.

In the summer of 1889, Tesla returned to New York, hardly able to wait to get back to his lab and start his studies. He could now afford the materials, and, as his own boss, was able to follow his own interests.

Resonance and frequency were playing on his mind, and, in his mind's eye, he constantly re-lived the scene of the snowball rolling down the slope, getting bigger and bigger. In that moment, he had witnessed a very small force causing a very big effect. Understanding why, was the problem. If he could grasp the principle, then he could use small forces to cause big effects. He tried thinking about the problem as a series of dominoes. Imagining a long line of these standing on end, close to each other, he pushed the end one over and watched all the others fall down along the row. The small movement of one domino could make hundreds fall down. How could such energy come from one small push? Tesla suddenly realized that the first domino was using the force of gravity to make all the other dominoes fall over, and that, in using this force, it could achieve much more than it could on its own.

What would happen if he used a series of dominoes, with each one slightly larger than the one before? With a long enough line-up, having pushed over the first tiny domino, the last giant domino could be big enough to crush the Empire State Building! Each link in the chain would increase the amount of energy available so, with a long enough chain, a tiny trigger would be able to topple an enormous block.

He started by looking at mechanical vibration, building a vibrating platform to test how things reacted at different speeds of vibration. This platform, as we shall appreciate in a couple of paragraphs' time, had strange yet dramatic effects on humans — effects that were potentially very messy!

All this was taking place during the period when Tesla, by now the most fashionable host in New York, was hosting sought-after dinner parties. One night the writer

Samuel Clements, better known under his pen-name Mark Twain, arrived with a group of reporters. As a child, Tesla had caught malaria and, as he lay sick in bed, had read *The Adventures of Tom Sawyer* and its sequel *The Adventures of Huckleberry Finn*. The exploits of Tom and Huck had inspired him so much that he was convinced that the books had restored his morale and made him well. Now, twenty-five years later, he was thrilled to meet Clements, and told him that his books had saved his life. Clements, subsequently, became a regular visitor to Tesla's laboratory at 35 South Fifth Avenue which was quite close to his home. Tesla, it seemed, wanted a father-substitute figure to publicly approve of his work; and Clements a chance to promote himself by associating with Tesla.

During one of his visits, Clements accidentally demon-strated the embarrassing consequences of mechanical vibration. By this time, Tesla had built a trembling platform to see how physical constants changed with the speed of vibration. The platform was mounted on elastic cushions and driven by compressed air so that it vibrated with great precision over a wide range of speeds. Tesla said that it was so accurate it could work like a precision clock pendulum. One day, while carrying out an experiment, he stepped on the platform when it was in motion and noticed a 'strange but pleasant' feeling as his body vibrated. He then encouraged his assistant to try the platform and established that he also found the sensation pleasant. Tesla, who had spent longer on the platform than his assistant, then discovered that the vibration had loosened his bowels to such an extent that he only just reached the lavatory in time to avert disaster. As he sat relieving himself, the truth dawned on him that he had invented a mechanical laxative.

Both he and his assistant, who were in the habit of rushing lunch in order to get back to work, started to use the platform, which also offered digestive relief, on a regular basis. As Tesla said at the time:

'We suffered from dyspepsia and various stomach troubles, biliousness, constipation, flatulence and other disturbances, all the natural results of such irregular habits. But after only a week of application [of the vibrating platform] all these forms of sickness disappeared as if by enchantment.'

As Samuel Clements suffered from constipation, Tesla suggested that his health would be improved if he tried 'mechanical therapy'. When Clements did so, he found the vibration so pleasant that he stayed on it longer than was prudent. Tesla tried to warn him that he had vibrated for long enough, but Clements persisted just a little longer. As a result, they had to send round to Mr Clements' house for a clean pair of white linen trousers. His constipation, though, was cured!

Greatly impressed with the device, Clements discussed its potential uses with Tesla who agreed that it had great medical benefits, and also believed it could become a popular beauty aid:

The greatest benefit will derived from my machine by women who will be able to reduce without the usual

> *tantalising abstinence, privation, sacrifice of time and money and torture they have to endure. They will improve much in appearance, acquire clear eyes and complexions and it may be safely predicted that long continued treatment will bring forth feminine beauty never seen before.*

It may have been this odd view on the essentials of feminine beauty that helped Tesla to become the kind of man who was not considered 'attractive to women'. Most men would hesitate before suggesting to any woman that she would be more attractive if she made copious use of a mechanical laxative machine! Tesla, however, had no such inhibitions and published his view that feminine beauty had a direct relationship to free bowel movements. Although handsome enough and certainly fastidiously clean, he never married or even had romantic liaisons. In fact, he never seems to have developed any interests outside his work.

The initial tests on a small vibrating engine almost caused a major incident when he strapped it to one of the iron column supports of his workshop and ran a range of vibration frequencies through it. The lab was on the upper floor of a four-storey building, and, as he increased the rate of vibration, he struck a speed that caused the building to quake in sympathy. The whole laboratory shook like a wine glass about to shatter when a singer hits the right note, and he had to turn off the motor quickly before the building shook itself to pieces. He had discovered that all objects have a natural frequency of vibration, called the resonant frequency; and that if you shake anything at this frequency the object will shiver wildly and eventually fall to pieces. Just as the snowball had become unstoppable when it went over its size limit as he rolled it down the slope, so had the small impulses of the vibrator grown until the whole building shook as if it were in an earthquake.

Tesla, who was now visualizing an earthquake machine, worked out the frequency of vibration that would be needed to shake the earth to pieces. Setting off a long sequence of small explosions, timed to match the resonant frequency of the earth, would split it into pieces, he declared. The time between the explosions would need to be about 1 hour 45 minutes. Each small explosion would shake the earth and the next would be added to the shock-wave at just the right time to cause maximum effect. Thankfully, he decided not to try this experiment, writing in his notebook that if he succeeded it would not be 'a desirable outcome'. He called this work his science of telegeodynamics, and said that it could be used for finding ores underground. In this, he foresaw the modern science of seismology that is now used by mining companies to locate seams of ore underground.

Also thinking about electric vibrations, it occurred to him that if he used a trembling contact he could make a very fast changing current. So, using high-speed magnetic switches and alternators, he made electric currents of many different frequencies. Now he could play with a whole range of currents from a single cycle per second up to 20,000 cycles per second.

When he used these high-speed currents in transformers, he discovered that he could make extremely high voltages of millions of volts. But there seemed to be a limit to what a mechanical switch could do, and, as a result, these mechanical ways of making high-frequency currents did not always give the accuracy he wanted. To overcome this, Tesla decided to make a vibrator with no moving parts, and ended up connecting a sort of Leyden jar, a capacitor, across a coil and found that together they could resonate millions of times a second. This electric circuit, called a tuning circuit, is the basis of all the radio and television sets we use today.

The tuning circuit makes radio possible. Tesla built a large number of different coils and capacitors and discovered that, as he changed the frequency of the currents he generated, different circuits would react. Out of a row of 100 different circuits, he could adjust his equipment so that just one responded. This happened without the need to connect wires to the circuits, so he now knew that wireless power could be transmitted.

He also developed a new type of wireless light bulb when he discovered that if he took most of the air out of a glass tube and stood it in a high-frequency electric field, it would glow brightly without the need to connect wires to the tube. He used these 'wireless' lamps to provide mobile light within his laboratory.

Tesla was not afraid of electricity. He knew that it

could kill – the electric chair was a fact – but he also knew that electricity passing through a human body was not always lethal. It was during this period that he first started passing currents through his own body to see what would happen. He knew that the effects of his mechanical laxative machine had been useful and now reasoned that, perhaps, electricity had other healing possibilities. He called the machine that he built after these experiments a Therapeutic Electrotherapy Machine.

For a while, from 1893-95, life was wonderful – a playtime for Tesla. He had money, a laboratory and no responsibilities except to his imagination. It was an incredibly creative period of his life. He proved the idea of broadcast radio before Marconi had even sent a one-to-one wireless message. He invented cordless fluorescent lights and used them to light his own laboratory. While experimenting with photographic lighting, he made a light tube which gave off X-rays and took photographs of the bones of his living hand. Years later, when he read Wilhelm Conrad Roentgen's paper on X-rays, he realized what he had made. He corresponded with Roentgen and sent him details of the experiments he had conducted, but he never tried to claim precedence over Roentgen for the discovery because he had never published his findings.

He invented the tuned circuit, and made any single circuit out of a bank of 200 respond to his command. He invented the cathode ray tube and the electron microscope before anybody else knew that electrons existed.

He invented the 'Tesla' coil to create enormously high voltages.

He dined at Delmonico's restaurant every night and afterwards took his socialite acquaintances to his laboratory to see his new inventions. He gave learned papers in New York, London and Paris, and was acclaimed a genius by his fellow scientists. He worked extremely long hours, seven days a week, until he fell asleep at his bench. He was gloriously happy. He spent endless sums of money on his experiments, but didn't earn any more.

After his mother's funeral, he had collapsed and temporarily lost his memory. But during this time he had been using his Therapeutic Electrotherapy device to make himself sleep, and probably caused his own amnesia with these experiments. Whatever the cause, he had recuperated by staying with friends in Watford, England, and reading the Book of Revelation. Inspired by his reading, he had then returned to New York with yet another new idea that he called teleautomatics (automatic control at a distance). And this is how he described it to his assistant:

> I will build a system for sending messages through the earth without wires. I may also be able to transmit electric force in the same way. I must first ascertain exactly how many vibrations to the second are caused by disturbing the mass of electricity which the earth contains. My transmitter must vibrate as often as the earth itself to put it in accord with the electricity of the earth.

As Tesla had not yet spent all the money he received from Westinghouse for his royalties, he then used what remained to build his first radio transmitter and receiver – a system that was way ahead of its time. Using a low-pressure air lamp as a sensor, the principle of the lamp was exactly the same as that for the thermionic valve or vacuum tube. Subsequently, valves such as this would be used in every radio made in the early twentieth century. Connecting the air lamp to one resonant circuit, he then sent current to a similar resonant circuit on the other side of his lab. The lamp lit up. He tried and perfected this idea within his laboratory, and then took the idea even further by running a small motor with wireless power instead of lighting a lamp.

In 1893, he gave a lecture to the National Electric Light Association in St Louis. Fresh from his success at the World Fair, he wanted to tell everybody about his idea for a world radio system. He opened his lecture with a few words about resonance: 'In connection with resonance effects and the problems of the transmission of energy, I would like to say a few words about the transmission of intelligible signals, or perhaps, even power, to any distance without the use of wires.'

He went on to say that he was convinced that such things were possible and that he intended to demonstrate it to the world very soon. He explained that the earth itself was a giant conductor which carried electric charge. To use it for transmitting would involve measuring its electrical capacity, something he had not yet managed to do, but he was convinced that it was electrified and that its natural electricity could be used: 'If I can ever ascertain at what period the Earth's charge, when disturbed, oscillates, with respect to an opposite charged system or known circuit, I will know a fact of the greatest importance to the welfare of the human race.'

He explained how a radio system would need to be connected to the earth and suggested that city water mains might be used for that connection. The radio would also need to be connected to the free space surrounding the earth, using an insulated high wire. He added that if a high-frequency current was connected between these two points, the effects it caused would be detected at great distances, if a suitable resonant circuit and air lamp were used. He finished his lecture with: 'This experiment will be of great scientific interest and will probably best succeed on a ship at sea . . . in this manner intelligence will be transmitted.'

In this public lecture, given three years before Marconi had even carried out his first experiments in wireless, Tesla had described the five basic features of a radio set-up which were: an aerial or antenna; a ground connection; an aerial-ground circuit for tuning; sending and receiving set tuned to each other's resonance or frequency; and an electronic detector of signals.

Tesla, in fact, had invented radio three years before Marconi, by discovering that if he generated a high-frequency electric current and passed it through a coil and a capacitor he could produce a resonance effect. This effect would then work at distance, without needing wires, and was a 'wireless' transmitter.

The cycle of events he discovered goes like this. When the current is first applied to the coil-capacitor circuit, the capacitor is not charged. All the current is sucked into the empty capacitor to charge it up. No current flows to the coil because the uncharged capacitor looks like a short circuit and takes in all the available electrons to charge it up. Once the capacitor is charged up, the current can start to flow to the coil. As it does, it makes a magnetic field around the turns of the coil, as well as an electric field along the length of the coil. This combined electro-magnetic field is a radio wave that travels off in all directions. Imagine the waves like ripples that spread out on a pond when you throw in a stone.

By the time the current in the coil has built up, the alternating current that was feeding the circuit will be starting to die away, getting ready to reverse its direction for a new cycle. As the supply current dies away in the coil, the stored charge in the capacitor starts to flow. This discharges the capacitor into the coil and keeps the electro-magnetic field in the coil going for much longer. Finally the capacitor will discharge and the whole cycle will start again.

The speed at which this happens depends on the number of turns, and the size of the coil, as well as the size of the capacitor. Different pairs of coil and capacitor will respond at different frequencies.

What Tesla had discovered was that if his supply current frequency was timed so that it started a new cycle just as the capacitor discharged, then the electro-magnetic field became very big. However, if the supply current reversed before the capacitor had discharged, the current still flowing from the capacitor to the coil over-rode the supply current and no radio wave was made. The circuit only made a radio wave if the supply current was set to the natural frequency of the coil-capacitor pair.

At the receiving end of his system, the electro-magnetic radio waves that were transmitted affected the coils. Using the analogy of a wave spreading out on the surface of a pond, when the radio waves reached the receiving coil, they made it bob up and down in time with the ripples. As the electric and magnetic fields passed through the turns of the receiving coils, they made electric currents flow in the coil. These currents rose and fell in time with the transmitted radio wave.

As the radio-induced current in the coil built up, it would start to charge the capacitor. When it died away, the capacitor could discharge back into the coil. If the natural resonance frequency of the coil and capacitor matched the transmitter frequency, the current from the capacitor worked with the coil and magnified the current. If the natural frequency did not coincide, the transmitter current was over-ridden by the capacitor current and they cancelled each other out.

Tesla made lots of different sized pairs of coil and capacitor. By changing the frequency of his transmitter, he could use the resonance effect to select any separate receiver to pass current to a light bulb.

This radio system, as mentioned earlier, had all the basic components of today's broadcasting systems, even supporting lots of separate radio channels. Although Tesla patented this system, he never exploited it commercially or even bothered to explain it to the world. It was not until six months after his death that his claim to have patented radio before Marconi was finally proved in an American court, but by that time both Tesla and Marconi were dead and all the history books continued to tell how Marconi had invented radio.

Why is it that Tesla is not remembered as the inventor of radio? Part of the reason was his perfectionism and fear of failure. He would not demonstrate any invention until he was sure that it would work perfectly. He was like an artist constantly fiddling with an almost finished painting, and never quite believing that anything was ready to show. When he spoke at St Louis, he could easily have demonstrated his radio working in the lecture theatre, but, instead, talked about a wireless system that worked over the whole world. As he was not ready to show that, he showed nothing, and spoke in vague generalities. He wanted to build and test his prototype, and secure his patent, before he explained more about his ideas.

Back in New York, he built lots of sending and receiving sets, tested these first within his laboratory, then took them out into the city to try them out. He refined and tested his equipment until at last he was ready. He made arrangements to hire a small ship on the Hudson river for the morning of 14 March 1895 and, as he returned to his hotel bedroom the night before the final test, must have been overwhelmed with excitement.

He certainly didn't sleep that night, not because of the excitement, but because he was awakened in the early hours by the police.

A fire had broken out in the lower floors of the warehouse below his laboratory. The building was well ablaze before anybody noticed and was entirely destroyed. All Tesla's equipment, materials, tools and records were lost. That was bad enough, but in his typical un-businesslike way he had not insured the property. His company had lost everything, just as he had sunk all the money he had got from Westinghouse into his radio experiments. He was ruined. All he had left were a few small royalties from the German use of his AC system. The money from those would feed him, but would not replace his burnt apparatus. Fortunately, he had a good memory so his ideas were not completely lost. At least he now knew how to build a radio system, but he had no lab, no tools and no money.

The *New York Sun* reported the fire, saying it was much more than a private calamity, describing it as 'a misfortune to the whole world'. The magazine *Scientific American* said that Tesla's loss 'could not be measured in dollars and cents', and expressed concern about the effect of the loss on his health and hoped a way could be found to help the man they described as 'the controlling engineer in the Niagara Falls Power Project'.

A rescuer did come forward, one who at first sight seemed highly unlikely: Edward Adams, the banker who worked for JP Morgan's banking group, and he gave Tesla a grant of $40,000 to help him get started again. Why did Adams do this? He seems to have taken a personal interest in Tesla and his work, and it has been suggested that he wanted Tesla to form a new company which would employ his son and be financed by the JP Morgan group. As JP Morgan controlled three quarters of the US electricity industry and had a strong influence over Westinghouse who controlled the rest, this would have been a powerful link for Tesla.

Adams, you will remember, had tried once before to bring together all the important patents in the electricity industry when he had attempted to set up a meeting between Edison and Westinghouse. He probably believed that Tesla was about to move the science of electricity forward again into a new wireless age, and it would have been in character for him to try to bring Tesla's new ideas into the General Electric fold or at least under the control of the JP Morgan bank.

The grant was also an excellent publicity move, coming as it did just after Westinghouse had switched on the Niagara Power Station. It showed the Morgan bank in a caring light and linked their name with Tesla, who, in press reports of the fire, was always described as the engineer who harnessed Niagara. If Adams could have used Tesla's bad luck to tie him permanently to the Morgan group, it would have been a tremendous coup. At the same time, Tesla would have been commercially focused and guided in his work, and adequately funded to carry on, at least with those parts of his work that fitted in with General Electric's plans.

Tesla, however, could only remember his last period

working for an industrialist. He had not enjoyed working at Westinghouse's Pittsburgh factories, and the heady thrill of six years' freedom was fresh in his mind. Like a child playing outside on a long summer evening, he could hear somebody calling him to come inside, but didn't want his dallying in the sunshine to end. There were more butterflies to chase, more flowers to pick, more discoveries to unearth. Even though night was coming, he wanted to stay out. He did. He took the grant, made suitably grateful remarks to the press as payment, and said 'no' to any further involvement with the Morgan group. He wanted to work on alone.

Tesla used the $40,000 grant to set up a new company in a new lab at 46 East Houston Street. Within four months of moving into his new premises, he started to rebuild the radio system he had lost in the fire. But, even with his photographic memory, it took him two years to get it working again and he did not receive the patents for it until 2 September, 1897.

Once the patent was secure, he staged an impressive demonstration of his invention in Madison Square Gardens, paid for by one of the mining engineers from Colorado who sent him a gift of $10,000 to help him get on his feet again. He showed a boat that was remote-controlled by radio. For the show, he had constructed a huge tank of water in the middle of the hall. The boat was about 5 feet (1.5 m) long and had a series of different coloured lights attached to it. Tesla made the boat follow his command, directing it around the tank, stopping and starting it, and making its lights turn on and off at his will. He allowed members of the audience to ask for the boat to move in a particular way, and then made it do it.

He even made the boat submerge and guided its underwater actions. At this moment, he had, in fact, demonstrated a formidable weapon – a remote-controlled torpedo boat – and described how his method of teleautomatics could be used to create a submarine destroyer: 'I am now prepared to announce my invention of a submarine torpedo boat that I am confident will be the greatest weapon of the navy from this time on.'

He explained that the torpedo boats the navy had used in the recent Spanish war were useless – too frail and too easy a target. He proposed a remote-controlled submarine that would carry six torpedoes and no crew. Because it did not carry men, it could be much more compact and could be operated as a submarine destroyer by men aboard a normal warship; it could even be sailed among the enemy's ships before firing its torpedoes at close range. He had even designed an automatic ballast-correction control that could re-trim the boat as it fired a torpedo. He estimated the cost of each of these re-usable submarines as $50,000, and said: 'It will have, also, the incalculable advantage of being absolutely invisible to the enemy, and have no human lives to risk or steam boilers to blow up and destroy itself.' He then added: 'How such a submarine should actually be used in war, I leave for the naval tacticians to determine.'

The navy never took up this weapon. If it had, the history of radio and warfare might have been very different. But, once again, Tesla had demonstrated his lack of commercial understanding, by confusing the people present on this occasion with two concepts. Having invented a multi-channel wireless telegraph, the first of its kind in the world, he had seen a way of using it to improve an existing naval weapon, and had only demonstrated a new type of torpedo boat. As a result, most of the people present failed to see the possibilities in the radio system he had used to control it. Tesla then added to the confusion by suggesting to his audience that he was controlling the boat by the power of his mind. Such misguided showmanship only confused the navy and the public further.

The resulting popular newspaper reports of Tesla's demonstration of 'mind control' certainly didn't do his scientific reputation any good. The navy didn't want to get involved with 'mind-controlled' destroyers and both it and the general public never realized that they had been offered a multi-channel broadcasting system. The result was that both inventions were ignored.

The poor response caused Tesla to lose interest in the commercial development of teleautomatics. Instead, he turned his attention to something new – the idea he had mentioned at his after-dinner speech in Buffalo – a system to transmit electric power without wires. He knew this would work because his remote-controlled boat had proved it. Now, he would show the world he was right. But to do so, he would need to try out some bigger experiments in a bigger laboratory.

chapter 10

The God of Lightning

When gloomy darkness hides the sea
And one no star and moon can see
They turn on the needle the light
Then from the straying they have no fright
For the needle points to the star.
  Guyot de Provins (translated by Nikola Tesla)

There are not many portraits of Nikola Tesla in existence. The editor of his 'autobiography' only managed to find four, but, viewed in sequence, they portray his rise and fall. His 'autobiography' was actually made up from a series of articles that Tesla wrote for the *Electrical Experimenter* magazine under the title 'My Inventions'. He always said he was going to write a 'proper' autobiography when he had spare time, but died before he got around to it. So, the compilation of articles and his working notes are the only documents he wrote explaining his thoughts.

The earliest photograph shows Tesla, with clean-shaven face, close-cropped hair, prominent ears and nervous expression, wearing a large floppy bow tie, just after his graduation in Prague. The next photograph, taken while he was working for Edison soon after his arrival in the United States, shows a confident young man who has taken to parting his hair in the centre and now sports a moustache above his half-smiling lips. His eyes gaze out at the world with the look of a man who knows for certain he is right. He is wearing a fashionable pinstripe suit and light tie; and shows the world an image of a cultured young man who can translate medieval poetry, quote literature and make a magnetic field rotate. His vulnerability and need for approval from parent-substitutes is well hidden in this picture.

The next photograph dates from just before the opening of Niagara Power station and the disastrous fire at his lab. A confident and successful man stares at the camera, head tilted like a bird of prey watching for a chance to strike. By now, Tesla had tasted fame and acclamation, and had taken to wearing sombre black suits more in keeping with his successful status. Even approaching the age of forty, his hair and moustache show no sign of thinning or losing their jet blackness.

The final photograph in the series, taken to celebrate the award of the Edison Medal, shows a bitter man. At sixty years of age, his hair is still thick and black, but his moustache is thinner. His mouth no longer has the enigmatic smile of confidence, but droops almost in a sneer. He is still wearing the same kind of formal black suit he wore twenty years earlier. It is easy to visualize this man quarrelling violently – as he did – with the equally irascible JP Morgan.

By 1899, Tesla was short of money but his technical skill was growing apace. He had still to make his most radical technical discoveries, and was dreaming of new success and telling people how he would soon be a multimillionaire. In technical work, he was a perfectionist. Scathing about engineers who rushed to sell their inventions before they had fully tested them, he said: 'An agreeable contrast is afforded by those who patiently investigate, contented to lose the credit for advances made rather than to present them to the world in an imperfect state, who form their opinions conscientiously, after a long and careful study, and have little to correct afterwards.'

Perhaps if he had been prepared to 'correct afterwards', he would have made more money.

He had proved that wireless could work, but was interested in more than simple telegraphy. He wanted to transmit power without wires and foresaw a global broadcasting scheme that would carry speech, music, pictures and electric power. What he had, was already far ahead of any work of Marconi's but it was nowhere near enough for Tesla the perfectionist.

Three consuming interests drove him. Inspired by the immense power of electrical storms, he wanted to investigate extremely high-voltage pressures of millions of volts. He wanted to know what happened if he made currents of thousands of amperes; and wanted to control 'the mysterious actions at a distance' caused by powerful electrical vibrations. He described himself as 'an exhausted wanderer in search of refreshing berries'. He was longing for more discoveries and wanted to be the first to travel new roads. In none of his notes did he show any interest in business affairs or who was going to pay for his expensive tests – tests that were the steps that would give electrical energy freely to the whole world. Once he achieved this goal, his place in history would be assured and his need for approval fully satisfied.

His mind was already picturing his next experiments. He was trying to go higher in both voltage and current than anybody had ever been before, and as he built bigger and more powerful equipment he realized that his Houston Street lab was not the place for this type of work. Indeed, he almost succeeded in burning it down when he created a spark of 4,000,000 volts that, instead of jumping to the terminal he had built, made a spectacular leap to the steel frame of the building. This decided him that he needed a bigger lab in a more remote area.

He had remained on friendly terms with his first patent attorney, Leonard Curtis, who had retired from the hustle of a New York legal practice to live in Colorado Springs where he had become a director of the Colorado Springs Power Company. Tesla's AC system had saved the local mining industry and, over time, his continuing success had made him a folk hero to the people of Colorado. When Tesla told Curtis that he needed to find a more remote laboratory, Curtis offered him, free of charge, a plot of land and as much off-peak electricity as he could use. In return, Tesla agreed to use the legal services of Hall, Preston, Bobbitt and Curtis of Little London, Colorado Springs, for any future patent work.

Curtis's offer was useful, but Tesla still didn't have enough money to set up a new lab in Colorado and had to borrow another $40,000 from friendly well-wishers. These included the owner of a restaurant he often frequented and the proprietor of his local hardware shop. Now, with enough money to get started, he left his New York lab in charge of his assistant, George Scherff, and set off for Colorado Springs, arriving on 18 May 1899.

By June, Tesla had finished building a large wooden shed to house the new lab and started to move in his equipment. Fortunately, during his stay in Colorado he kept a detailed notebook. He didn't often keep notes, preferring to trust his memory, but the problems of running two labs must have convinced him that written instructions were essential if George Scherff, back at Houston Street, was to know exactly what to make for the new tests. Tesla's engineering diary runs from 1 June 1899 to 7 January 1900, and gives the best picture we have of his way of working, and of the thought processes he went through as he developed his ideas.

Some writers have suggested that he had learnt something from the loss of his laboratory in Fifth Avenue and had started to keep records, but this seems unlikely because he stopped keeping a working diary when he returned to New York and no longer had the problem of working between two sites.

The Colorado Spring Notebooks, which fortunately were published in their entirety by the Tesla Museum of Belgrade, are a valuable insight into the strange wonderful mind of the man himself. They consist mainly of a series of notes about experiments and the results of those experiments, but interspersed with the technical information are little cameo descriptions of the surrounding scenery or wildlife. He discourses on the effects of mountain climates on health and nature, of thunderstorms and clouds, but says little about the domestic details of his work. For example, he took with him an assistant, Mr Alley, who figures in many of the photographs and is named in the captions, but he is never mentioned in the notes. Tesla never seemed to show any great interest in people.

In the notes, he writes about the three main goals he had set himself when he went to Colorado Springs. These were: to develop a transmitter of great power; to perfect means of individualizing and isolating the energy transmitted; and to ascertain the laws of propagation of currents through the earth and its atmosphere.

The early pages of his journal reveal him wondering how to approach his problems. He makes lists of possible techniques and their applications. He is clearly worried about how to carry out measurements and about the types of energy he needs to detect. Thinking about the

Tesla's laboratory and tower at Colorado Springs.

THE GOD OF LIGHTNING

problems of measurement took him down a road that could have been extremely commercial if he had thought to sell the instruments he designed. But he didn't.

The mountains of Colorado were an ideal place for him to study natural lightning. The purity of the atmosphere and the high vantage point of the site gave magnificent views. Mountain ranges up to 150 miles (240 km) away could be seen clearly. The air was so dry in those high mountains because there was no water mist to obscure vision. The dryness of the mountain air and the stillness of the surroundings meant that sounds would carry for tremendously long distances. Tesla noted that when a bell rang in the city (some miles distant) it seemed as if it were ringing by the very door of his laboratory.

The countryside surrounding the lab was barren with very little vegetable or animal life – just a few prairie dogs lived in the desert-like conditions – and Tesla bemoans the fact there were no birds to hear singing or to watch. The changing sky, however, was a constant source of interest. For his own amusement, he categorized the different type of clouds he saw: red clouds, seen in the early morning; white clouds, seen in the forenoon; clouds like lumps of gold, mainly seen at sunset; and clouds like lumps of incandescent metal, which he thought must be converting dark radiations of the sun into visible light.

He thought the Colorado light was far better than anything he had seen in Italy, and commented to Lawyer Curtis: 'We are used to speaking of "Sunny Italy" but compared to Colorado that country might be likened to foggy England.'

The components of Tesla's large magnifying transmitter
inside its protective ring.

He was surprised by the heat of the Colorado sunlight. Indeed, it took him so much by surprise that
his equipment was damaged because he carelessly left
it in the sunshine. A transformer and several barrels of
concentrated salt solution were melted and ruined by
the heat, and a wooden ball, coated with tinfoil, actually
caught fire.

He believed the Colorado sunlight was beneficial to
good health, as it was producing a 'specific germicidal
effect'. Germs were a constant fear of Tesla's. He was
obsessive about washing his hands, often doing this two
or three times during a single meal. And he made a

tremendous impression on the staff at the Alta Vista Hotel by his copious appetite for clean linen. He used a clean towel every time he dried his hands. (Ever since the periods of enforced squalor, when he had travelled across the Atlantic without his luggage and worked for a year digging ditches, he had vowed that he would never again use a towel twice.) This obsession with cleanliness also extended to his bed linen, and he insisted that this be changed every day when his fresh supply of eighteen towels was delivered.

His experiments were going well. He had built some very sensitive devices for measuring electric fields and, happening to have one of these set up during a thunder storm, was able to measure the effects of lightning discharges as the storm moved away from his lab. As the storm moved steadily towards the distant mountains Tesla saw, from his instruments, that the strength of its electrical discharges was growing to a peak and then declining before repeating the whole cycle. The lightning storm was producing something that a modern engineer would call a standing wave. This type of radio wave produces steady voltages that can be measured as you move away from its source, and the pattern it creates looks like a series of repeating hills as the voltage climbs to a peak value and then falls away to nothing before starting the whole cycle again.

Tesla realized that this effect proved that the earth and its atmosphere were charged with electricity. The storm was making a pattern on the fixed background charge of the earth. He would have been delighted by

Nikola Tesla, sitting inside a cage of artificial lighning.

observations, recorded in 1997 by the Russian Mir space station, that lightning storms always appear in regularly spaced lines across the dark side of the earth. He would have said: 'I realized that when I was in Colorado.'

This was an extremely important finding. It meant that the transmission of energy without wires was not just possible, but practical. These are his own words as he realized what his measurements were telling him:

> As the source of the disturbances [the storm] moved away the receiving circuit came successively upon their nodes and loops. Impossible as it seemed, this planet, despite its vast extent, behaved like a conductor of

limited dimensions. The tremendous significance of this fact in the transmission of energy by my system had already become quite clear to me. Not only was it practicable to send telegraphic messages to any distance without wires but also to impress on the entire globe the faint modulations of the human voice, far more still, to transmit power, in unlimited amounts to any terrestrial distance and almost without loss.

To help see why this is so important, imagine a water trough. If there is water in the trough and you move your hand slowly back and forward in it, the water starts to move. Move your hand to and fro quite slowly and the water will sway backwards and forwards building up waves. Keep doing it and the water will begin to slop over the side. This large movement of water is caused by the very small movements of your hand. If the earth did not have an electric charge, any attempt to vibrate the electric field near the earth's surface would not have worked. The earth itself would have soaked up too much electricity before any changes could happen. It would have been just like trying to make waves in an empty trough without the amplifying effect of moving water. Resonance happens because the trough is filled with water that magnifies the slight shaking of your hand. The earth allows standing waves to be manipulated by small amounts of energy because there is a lot of electric charge to move about and increase the small forces of the transmitter.

Tesla's notes shows his excitement as he writes:

> *Observations made last night. They were such as not to be easily forgotten . . . a wonderful and most interesting result from the scientific point of view. It clearly showed the existence of stationary waves for how could the observations be otherwise explained. This is of immense importance.*

He was now quite sure that he could use his knowledge of resonance to transmit power without wires. He had already discovered he could make the earth ring like a bell, with a stroke every two hours. Now he could also make it resonate electrically.

He found the electrical resonant frequency of the earth to be about ten cycles a second. (This was a remarkably accurate result as the value that engineers use today is about 7.8 cycles per second.)

His radio work was far more advanced than that of Hertz and Marconi, the other pioneers of wireless. They used much higher frequencies that did not resonate the earth. Tesla used very long wavelengths for his radio waves. They travelled easily right around the earth, and we now call these type of waves Very Low Frequency (VLF) waves. They have been used by navies to keep in radio contact with submarines no matter where they might be. VLF radio has the advantage that it can be received anywhere on the earth's surface or even beneath the waters of the sea. This use of Tesla's work, however, was not exploited until many years after his death.

Tesla's very first experiments succeeded in circling the earth with radio waves while Marconi's non-resonant short waves couldn't send a signal further than 62 miles (100 km). Tesla's designs were years ahead of Marconi's. If he had not carried out such 'a long and careful study . . . [leaving] little to correct afterwards', he could have made his fortune, but that was not his way. Instead, he fiddled with his system until Marconi had overtaken him commercially.

From the outside, the laboratory at Colorado Springs looked like a large wooden barn about 40 feet (12 m) high. It had large central doors with windows to either side and was braced with massive balks of timber on the other three sides. Above the entrance was a lattice-work mast, three times the height of the building. From the top of the mast ran a single copper pole, twice as high again, supporting a large copper ball on the top. The copper ball was over 200 feet (61 m) from the ground. Just above the lattice mast, a mushroom-shaped insulator had been mounted. The copper pole went down into the lab to the top of the largest and most powerful Tesla coil ever made. Tesla called this device his magnifying transformer and it was capable of generating voltages of 100,000,000 volts (100 megavolts).

Inside, the building looked even more like a large barn. The centre of the floor space was dominated by the gigantic coil of the magnifying transformer. The roof above could be opened to the sky. In the centre of the barn was what appeared at first glance to be the auction ring of a strange cattle market. The floor of the ring was made of wooden boards and around its perimeter was a 6 feet (2 m) high wooden slat fence with a strong wire cable running along the top.

Down from the roof opening came an upright wooden shaft, like the centre pole of a circus big top, into the centre of the ring. This stake ended in the middle of a large raised cage. Like a cross between an over-large auctioneer's podium and a cage for displaying wild animals, the structure was made of upright wooden bars about 9 feet (3 m) in height. Around its circular frame were wrapped many turns of wire to make a giant coil. The cage was supported on a spindly platform, made from balks of rough timber, and stood about 3 feet (1 m) above the floor of the ring. Around the ring stood what looked like a herd of surreal creatures: a vertical cylinder standing four-square on splayed legs; a giraffe-like structure with elegant legs and a long pole-like neck supporting a copper ball of a head; and two squat upright cylinders without legs standing in front of a tall thin gangly four-legged cylinder with a tuft of wire protruding from its head.

Outside the ring, standing like ranks of massed farmers at an auction, stood row upon row of squat square capacitors. They clustered close round the ring as if their terminal eyelets were trying to weigh up the strange creatures on parade. By the doorway stood a tall, fluted transformer, like a doorman in a striped uniform guarding the entrance.

The transmitter was capable of producing 10,000 watts of power, enough to run about 200 average power light bulbs. The transformer's windings were made in the shape of a pulled out spiral. It was designed to be pulsed with low-voltage AC current from the local power station and transform it into extremely high voltages with much higher frequencies. Many of Tesla's experiments at Colorado were carried out with frequencies of between 480 and 130,000 cycles per second. He was concerned to make sure that his connection to the air was a good one and, as he was trying to improve it, he discovered the principle of the resonant antenna that we use today. He found that an aerial of one quarter of the wavelength of the transmitter frequency produces the highest voltage at the radio set.

As he was setting up his gear, Tesla found that different speeds of vibration made standing waves of various types on his copper pole. He had discovered something that modern engineers call the wavelength of the signal. Tesla could change the length of his pole and, by careful 'tuning' of the length, could get maximum voltage at the copper ball on the end. The aerial on a modern mobile phone still uses the tuned aerial that Tesla invented. His notebooks reveal that he spent a long time adjusting his machines to generate the highest possible voltages, but at last he was ready to begin.

He had placed a number of receiving stations around the lab. Each consisted of a large coil with one end connected to a peg driven into the ground and the other connected to a gas glow-lamp. These looked just like old oil drums with light bulbs sticking out of the top. The outer ring was an earthing loop to prevent stray sparks from reaching the wooden building. The wire on top of the retaining fence was connected to the earth. Photographs of some of Tesla's tests show streamers of lightning running from the centre cage outwards to the earth ring-fence in a spectacular display of electric power.

On the wall by the transformer was a small wooden structure that looks like a dovecote. This is the cover over the main power switch for the magnifying transmitter. Tesla later put an additional switch inside the ring after having a near fatal accident when he was working alone in the laboratory one evening and accidentally started up the magnifier while he was inside the ring. A curtain of sparks crackled from the top of the centre cage outwards to the earthing ring, trapping him inside. Once started, the only way of stopping the lightning machine was to turn off the main power switch, which was on the other side of the curtain of lightning. Fortunately Tesla was wearing his thick-soled rubber boots. He reasoned that if he leapt through the lightning wall he would not be earthed through his rubber boots and so the electricity would not pass through him. He was right and lived to tell the tale, but he rearranged his lab to avoid the danger in the future. It was not an experiment he wanted to repeat.

Tesla warned his lawyer friend, Mr Curtis, that he would soon be making use of large amounts of current. He agreed with the power company to carry out his tests in the early evening when the city was not using much electricity and the generators had spare capacity. He knew that he could send electricity through the air to light up his receiving stations, but what he needed to find out was what would happen if he used very high voltages? He was about to try to make lightning.

The first test was very short. The big coil was connected to the town supply for just one second. The result was a spectacular spark from the copper ball. Tesla knew from his study of resonance that if he turned the power on for longer he would get bigger voltages. He stood outside where he could see the mast and the copper ball clearly. 'Keep it switched on until I tell you to stop,' he called out to his assistant.

The sparking increased, growing in strength, until the first bolt of Tesla-made lightning sprang down from the ball. The strike was over 200 feet (61 m) in length. There was a mighty clap of thunder as the electricity ripped the air apart. The local newspapers reported that the noise could be heard as far off as Cripple Creek, 20 miles (32 km) away. It was followed by a sudden silence as the power was cut off from the magnifying transformer. 'Turn it back on, I haven't finished yet,' Tesla said, his angry voice echoing through the stillness.

But Mr Alley, his assistant, hadn't turned off the power. The experiment had overloaded the local generator and it had caught fire. There were no more tests that night – indeed there were no more tests for quite a while. The power company decided it would use a standby generator to supply Tesla's lab so that he would not in future accidentally cut off the town. Curtis told Tesla that a standby generator would be available for him to use just as soon as he had repaired the one that had been burnt out.

The very first job that Tesla had done when he arrived in America twenty-six years earlier was to repair a generator. Once again his practical skills were brought into play and, within a week, he had rewired the generator so the tests could continue.

Over the next few weeks he worked out exactly how to transmit electrical power without using wires. He tested receiving units in various parts of the town and sited them at greater and greater distances. The maximum range he managed during these tests was 26 miles (42 km) and he succeeded in running 10,000 watts of light bulbs from that receiver. Two-hundred lights, running from one of his oil-drum receivers, were an impressive sight. It was a major proof of his wireless power system and gave Curtis more patent work. That first patent for the wireless transmission of electrical power (No. 645 576) was issued to Tesla on 20 March 1900. By 1902, Tesla would hold every patent controlling wireless transmission of power.

Once he had proved his wireless power transmission method, Tesla used the lab to explore how lightning worked. In this, he was following an old American tradition started by another famous American, Benjamin Franklin. Franklin became interested in lightning when he was Deputy Postmaster General of Philadelphia and carried out many tests to understand what lightning really was. Lightning, caused by thunder storms is, as we know, extremely destructive if it strikes a building.

Franklin believed that lightning is natural electricity and, in the midst of a thunderstorm, carried out a very dangerous experiment with a kite to prove it. His study helped him make the lightning conductor: a small metal rod placed on the top of a building and connected to the ground by a thick wire. This simple device protects buildings from lightning and has saved many lives since Franklin invented it. Tesla, however, wanted to go further – he wanted to make a lightning machine.

He was fascinated by ball lightning, but had never seen it. By boosting the power of his lightning machine he succeeded in making fireballs. A fireball is a ball of light that sometimes manifests during a thunderstorm. They are about 25 centimetres (10 inches) in diameter and can be all sorts of colours. Ball lightning does not behave like forked lightning, but moves very slowly and sometimes seems to hover in the air.

Tesla already knew how to make gas light up inside a tube, had been making glow tube lamps for many years, and was using these as sensors to test if he was transmitting electricity. A neon glow tube works by making something physicists call a plasma. Plasmas are made when some electrons are pulled away from the gas molecules by a strong electric field to form ions. This mixture of electrons and ions, now called an ionized gas, will glow brightly and give off light and heat. The most common example of a plasma giving off light and heat is the sun. The sun is a giant plasma fireball that heats our planet. The fireballs made during lightning storms are miniature suns, and Tesla wanted to know how to make them.

Modern physics believes that fireballs are plasma zones caused by electric current flows. They are made up of a strong electro-magnet field that holds ionized air within it. (Ionized air has molecules that have had some electrons taken away, making it electrically charged.)

The balls are caused by a resonance effect with the earth's electro-magnetic field. To create a fireball needs a lot of energy. To form one of about 35 centimetres (14 inches) diameter takes at least 5,000 watts. Once the plasma is triggered it will resonate with the earth's electric and magnetic fields and sustain itself for many seconds.

Tesla's understanding of the way fireballs are made was remarkably perceptive. He was well ahead of the physics of his day, and his explanation still makes sense to modern physicists. From the theoretical comments in his Colorado notes, it is quite clear that he understood how to create electro-magnetic plasmas seventy years before the term was even coined. His patents were named as the source of the idea of a plasma shell weapon for zapping spy satellites which was built, tested and shown to work as part of the Star Wars project.

By the middle of January 1900, Tesla was broke again. He had used up the $40,000 loan and needed more money to continue his work. He had made a series of extremely important discoveries. He could transmit telegraphy signals worldwide, at a time when Marconi was being applauded for struggling to transmit messages over very short distances. He could transmit electricity without wires and with virtually no loss of energy. Because he didn't use wires there was nothing to heat up and waste power. He could make artificial lightning and understood the awesome power of plasma fireballs. He also knew how to make the earth's electrical field resonate in ways that he believed could control the weather.

These discoveries were extremely important. Many scientists believe that ball lightning may well be the key to understanding the processes that make the sun work, and to using these to make cheap electricity. Our present-day nuclear power stations use a process called fission, which makes radio-active waste materials, such as plutonium. Thermo-nuclear fusion holds within it a possibility for cheap and clean nuclear power. It works by turning hydrogen into helium and gives off energy as it does so.

Fusion, however, takes place at such high temperatures that there is no known substance on the earth that will not melt if we try to contain the process. The only possible thing that could hold such hot materials is a plasma bottle. This is exactly what Tesla made when he produced his ball lightning, and the discovery had vital implications for the study of quantum physics. This brought Tesla into verbal conflict with Einstein in later years because these experiments convinced Tesla that gravity was a field effect which of course did not fit with Einstein's idea of gravity as curved space.

The other important discovery that Tesla made in

Colorado was how to create electrical standing waves to transmit power around the world. Today's scientists know that there is an area in the earth's atmosphere called the Schumann Cavity, the space between the ground plane of the earth's surface and the electrically-charged ionosphere. Electrically, the Schumann Cavity looks like a world-wide capacitor in the atmosphere. Tesla succeed in making this entire world-wide capacitor vibrate with power. Anybody, anywhere, if they had had the right sort of simple receiving equipment, could have drawn down the power he was transmitting at that time. It seems that this cavity affects the distribution of lightning storms throughout the world and Tesla had proved that he could make it resonate.

Although he was only interested in working for the betterment of mankind, he needed money and commercial help to develop his ideas. But, to find funds, he would have to go back to New York. He was well on the way to completing a set of patents covering all aspects of wireless transmission of power and intelligence. Convinced these patents were worth a fortune, he set off for New York believing he was about to make an major impact on the world and raise real money for his future experiments.

chapter II

Wireless Power

'A new and glorious age for humanity!'

The practical success of an idea, irrespective of its inherent merit, is dependent on the attitude of the contemporaries. If timely it is quickly adopted; if not, it is apt to fare like a sprout lured out of the ground by warm sunshine, only to be injured and retarded in its growth by the succeeding frost.

Nikola Tesla

The dream of free power for the world that Tesla had first shared with his mother during that long-ago thunderstorm was about to become reality. Just as he had harnessed the power of the Niagara Falls, so he had now harnessed the earth itself as a conductor of electric power, and brought this great discovery back with him from Colorado. But who wanted it?

He had developed a far better way of distributing power than his own AC system which, in itself, was vastly superior to Edison's. But, despite its virtues, the AC power system had only been successful because Tesla had been lucky enough to find in George Westinghouse a marketing champion who was looking for a way to break Edison's grip on electric power. For a short happy period, Tesla's interests had matched those of Westinghouse, and his AC patents had proved to be Westinghouse's tool for breaking Edison's singular hold on the electric power market. The time was right for AC power to succeed, and the words, quoted at the beginning of this chapter, written towards the end of Tesla's life, speak a truth that he himself was slow to learn.

As he arrived back in New York, his heart set upon becoming a major benefactor to mankind by bestowing his means of sending power anywhere in the world, for anyone to make free use of, how to achieve commercial gain from his work had not even entered his mind. Somebody, he believed, would put up the money for his idea. Driven by an insatiable curiosity about the world and its workings, and a childlike belief that he could improve it, he had already said as much during his protracted after-dinner speech in Buffalo when he had been honoured for his Niagara work: 'If we want to reduce poverty and misery . . . Power is our mainstay, the primary source of our many-sided energies. With sufficient power at our disposal we can satisfy most of our wants and offer a guarantee for safe and comfortable existence to all.'

Now Tesla was returning from Colorado with that comfortable existence for all and, this time, he thought he knew what he was doing. An old hand at developing new power technologies, he intended to repeat his first success exactly, and wrote Westinghouse a long rambling letter suggesting a fresh partnership to exploit his new system:

> *I have just returned from Colorado where I have been carrying out some experiments . . . The success has been greater than I anticipated . . . The demonstrations which I have made in Colorado are of such a nature that they preclude the possibility of failure . . . I have neglected to make provisions for money . . . Being compelled to borrow money I turn to you to ask if your company will advance against my royalty rights . . . or if preferable whether they would buy them outright.*
>
> *Needing, however, the money and thinking that my rights in your hands would if anything enhance their value since it would give your company an additional cause to push the business, I have ventured to make this suggestion.*

Westinghouse was doubtless alarmed by this letter. By 1900 his own company and General Electric had, between them, built up a total monopoly on AC electric power supply and both companies were doing very well out of the business. Now the message inherent in Tesla's letter was: I've invented something that will turn your previous investment into worthless scrap. I'm going to give electricity away instead of charging for it, but I need money to help me do this. How about lending me some? Not surprisingly, this time, the interests of Westinghouse and Tesla were incompatible, and Westinghouse turned down Tesla's extremely naive offer.

Still in need of money and, what's more, being pressed by Colorado creditors, Tesla's solution to his problems was to write for *Century Magazine*. The editor, an old friend, who knew that Tesla's name would sell his magazine was happy to commission an article about his exploits. But even in the simple task of writing about his exciting inventions, Tesla was not an easy man to manage. The article that he wanted to write was a philosophical discussion of the motives and aspirations of humanity, not the exciting account of what it felt like to make lightning that the magazine wanted. His first draft was returned with the comment, 'You are giving people Euclid and they don't want Euclid. They will say it is obscure and dull when it is only deep.'

Three times the article was returned for rewriting until a final compromise was reached, but the result didn't really satisfy either party. *Century Magazine* had to settle for a sensational article entitled 'The Problem of Increasing Human Energy' in which Tesla proposed a 'World System' to make possible 'the instantaneous and precise wireless transmission of any kind of signals, messages or characters to all parts of the world'. There would 'be no limits other than those imposed by the physical dimension of the Globe'. It was not the colourful telling of his taming of electricity that *Century* had asked for.

Writing about the mathematical laws that governed humanity's actions, Tesla rambled on discussing the problems of drink, claiming that impure water caused more deaths than whisky, advised against gambling and meat-eating, and made the startling claims that a crystal is a form of life, and that the invention of the aeroplane would bring about universal peace. He talked about sending heat to the North Pole, forming ice in the tropics, sending pictures round the globe and broadcasting music to the world. Most outrageous was his claim that man would be able to draw unlimited free electricity from the earth and labour would no longer be needed. Peace and prosperity would be universal. It was a vision of a pure Utopia with no economic insight to temper it, and was the foundation of Tesla's reputation as an outrageous prophet of the future.

Among the readers of 'The Problem of Increasing Human Energy' was John Piermont Morgan who remembered the author as the man who had spoken about wanting to create a wireless power system at the celebrations of the Niagara Falls power station. Now Tesla, who had tamed Niagara and had a track-record for producing ideas that worked, was claiming to have realized his dream. JP Morgan was interested and invited Tesla to dinner.

Tesla's 'World System' made use of five of his most

important inventions/discoveries which were: the Tesla coil (a device for making high-voltage, high-frequency currents of tremendous strength); the magnifying transmitter (a machine to generate lightning fields to resonate with the earth's own electric charge); the wireless system (Tesla's method of transmitting electric power without wires); the art of individualization (the means by which each receiver could be 'tuned' to its own individual wave length. Using the principles of resonance Tesla had designed a way of making sure each station could receive only its own separate messages); terrestrial stationary waves (Tesla had discovered that the earth would respond to electrical vibrations of a certain rate – it was like a giant tuning fork vibrating in sympathy with sounds of the right pitch – and planned to make stationary waves around the earth and use the earth's electric field to transmit electricity without any loss of power).

He outlined a radical vision of what his World System would give to Humanity, listing twelve world-wide major benefits that would result from connecting all the world's telegraph offices. This action would provide a secret and secure government message service; allow any telephone-user to speak to any other; provide a news transmission service for the world's newspapers; offer private citizens the opportunity to exchange private messages rapidly and safely; link all stock-markets; transmit music anywhere; offer a world-wide automatic system for setting clocks 'with astronomical precision'; allow for the transmission of typed or hand-written script or cheques; provide a navigation system that could tell any ship's navigator exactly where the ship was at any time; offer a means of printing out messages to any location on land or sea; allow for the reproduction by wireless of photographs or drawings; and make possible the transmission of electric power to any site in the world.

Reading this list almost 100 years after it was written, we will see nothing unusual about the first eleven items because they are all in regular use today, although linking the world's stock-markets and the system of navigation have only been put into effect in the last few years. For Tesla to get eleven out of twelve predictions exactly right, however, is good enough going for any prophet. Even his last prediction has been the subject of tests in recent years and has been shown to be a technical possibility.

When Tesla put forward these ideas, Marconi was still struggling to send Morse code messages a mere 50 miles (80 km). Yet it was Marconi who turned many of Tesla's ideas into a working reality and who is now remembered as the father of radio. The difference? Marconi had a head for business, Tesla did not. Marconi worked with governments and the military while he perfected his mechanisms, Tesla insisted on working alone until he had completely finished his system. Indeed, so strong was Tesla's drive to produce a perfect system that he turned down offers of money rather than allow the use equipment that he didn't think was ready. He refused an offer from Lloyds of London to buy the wireless system he had shown in New York, that they wished to fit to yachts taking part in an international race.

He told George Scherff, his long suffering assistant, that playing around with short-range dispatches between ships was a waste of time, and he couldn't lose good research time to make the equipment. However, if he had accepted the offer, it would have given him a ten-year lead over Marconi and he, rather than Marconi, would now be remembered as the inventor of radio. Scherff suggested to Tesla that he should build the first set of equipment, hire a manager to run a company selling marine wireless sets, and use the money that this would make to carry on with his researches. But Tesla couldn't be bothered and lost the chance. His failure to understand the ways of commerce had now reduced him in the eyes of the public to the status of writer of outrageous science fiction, and he even seemed to take a perverse delight in this, insisting that soon he would be sending messages to Mars. The fact that this proved possible, sixty years after his death, did nothing for his bank account.

Some journals accused Tesla of making up stories just to get his name in the newspapers. Rival scientists, writing in *Collier's Weekly*, said:

> *Mr Tesla's writing's must be judged with extreme caution, electrical experiments can only be judged by commercial success and Mr Tesla's speculations are so reckless as to lose interest and his philosophy so ignorant as to be worthless.*

But, when I was watching the delightful antics of NASA's 'Sojourner' being controlled from Houston over a radio link to Mars, I couldn't help thinking of Tesla. Seeing the images of 'Sojourner's' on-board camera transmitted live from the red planet to the world-wide Internet, would have pleased rather than surprised Tesla. His ghost must surely have been thinking 'I told you so!'

Writing yet another 'outrageous' article in *Collier's Weekly* on 9 February 1901, Tesla said:

> *The idea of communicating with other worlds is an old one. But for ages, it has been regarded merely as a poet's dream forever unrealisable. And yet, with the invention and perfection of the telescope and the ever-widening knowledge of the heavens, its hold on our imagination has been increased, and the scientific achievements during the latter part of the nineteenth century, together with tendency towards the nature ideal of Goethe, have intensified it to such a degree that it seems as if it were destined to become the domination idea of the century that has just begun.*

How right he was, but when he wrote those words most of his readers believed him to be wild and unworldly. There were exceptions, in particular JP Morgan who had got rich by looking ahead and spotting future trends. Morgan had been the driving force behind the formation of the General Electric Company and, in Tesla's 'World System', he recognized a threat to the comfortable duopoly of GE and Westinghouse. He knew that the success of both the companies was rooted in Tesla's AC power patents, and appreciated that Tesla was now proposing a new system to replace these. If Tesla were right, he would once again turn the electricity industry on its head; if he were wrong, Morgan's investment in GE would be safe. As lack of money seemed to be the only thing that was preventing Tesla from proving his new system, Morgan decided to back both horses to be certain of a win. It was, after all, much better to control a threat than worry about it.

In dire need of money, Tesla was in no position to bargain over the harsh deal that Morgan offered him – $150,000 for a fifty-one per cent share of all the 'wireless' technology patents that he might develop. Tesla, who saw the money as the only means of continuing his research and proving himself right, had no other way of funding the work – even Westinghouse had turned him down – so, once again, he sold for a pittance an invention that could have made him rich.

On 10 December 1900, Tesla wrote to Morgan confirming the deal:

> *The control is yours, the larger part is yours. As to my interest, you know the value of discoveries and artistic creations, your terms are mine.*

Part of the deal was a gentlemen's agreement not to discuss their terms publicly, an agreement that Tesla honoured right up to his death. But JP Morgan was less than a gentleman, and the full extent of his conquest of Tesla only became apparent long after the inventor's death when both their letters were deposited in the US Library of Congress.

Publicly supporting Tesla, in order to protect his own business interests, was a shrewd move on Morgan's part. It allowed him to ensure that Tesla was given insufficient funds to achieve success; and, having made sure that his 'support' for Tesla was widely reported in the newspapers, warned off any other potential investors. The result was that Tesla was completely at the mercy of a man who was only too aware that a controlling interest in the patents gave him as much right to suppress as to exploit them.

Tesla, of course, did not suspect this, he was far too busy dreaming dreams of providing free electric power for all and intent upon building a model city that would serve his new World System. In order to do this, he also struck a deal with a real estate developer who owned large tracts of land on Long Island where he was allowed to set up his lab and transmitting tower. The developer hoped to build and sell houses around the site to the hordes of people who would be employed at it and so make a good profit from the deal. On 23 July 1901, work started on the site that was to be called Wardencliff.

Tesla also managed to persuade a well-known architect, Stanford White, to design the buildings that would house the World System. These included an enormous wooden tower that he needed to support the electrodes of his magnifying transmitter. Any high tower is at risk from high winds of winter on an exposed site, but

White succeeded in designing a complex but stable structure that survived for many years as a testament to his architectural skill.

Two year's later, Tesla had a laboratory and power-house, built out of brick, with White's substantial wooden mast towering above them. All he needed now was a copper dome to be fixed to its top and he would be ready to transmit. Having spent almost $200,000 getting this far, he was still nowhere near ready to test the plant. He had sold most of his personal assets and taken out a bank loan of $10,000, but had again run out of money. On the basis of what he had already achieved, Tesla went back to Morgan for more money to complete the project. Morgan, for the financial self-interest reasons that we now understand, replied 'No'.

To try to persuade Morgan that the project was viable, Tesla spoke to the Canadian Government, which agreed to give him 10,000 horsepower of electricity for the next twenty years if he would agree to build a plant by the Falls to transmit wireless power to more remote parts of Canada. At this stage, Morgan started to show his hand more clearly. He wrote to Tesla refusing him any further money and made sure that this refusal got into the Press. Rumours started to spread that if JP Morgan was afraid to invest in Tesla, then the project must be flawed. Perhaps, it was thought, Tesla's critics were right after all: he was a wild dreamer who could not be relied upon to deliver a commercial success.

Reading through Tesla's letters to JP Morgan (micro-film archive in the US Library of Congress), it is clear that the inventor was getting increasingly desperate. On 14 January 1904, he wrote:

> *We start on a proposition . . . financially frail. You engage in impossible operations, you make me pay double, make me wait ten months for machinery. On top of this you produce a panic. When, after putting all I could scrape together, I come to show you that I have done the best that could be done you fire me out like an office boy and roar so that you are heard six blocks away; not a cent. It is spread all over town, I am discredited, the laughing stock of my enemies.*

The letters, covering a period of 10 December 1900 until 16 February 1906, reveal the closing of the trap that Morgan had set on the too trusting Tesla. Forced to accept that Morgan was not going to provide any more money for the completion of his World System, Tesla now tried to sell his only remaining asset to raise enough cash – himself. He circulated an advert, offering himself for hire as a consultant engineer. As a piece of advertising, it was impressive: printed on vellum, issued from the Waldorf Astoria Hotel, and running to four pages.

The front page was a personal statement from Tesla, headed with the Latin phrase '*Nihil in Sacculo quod non fuerit in Capite*' (Nothing is in the wallet that was not first in the head). The manifesto shows the Wardencliff tower under which is written 'Power transmission without wires,' and around the front sheet a hand-drawn border with the linked words:

Electrical Oscillation Activity Ten Million Horsepower – Oscillating high frequency coil – Tuned Circuits configured – Control of industrialised automatons at a distance – Sensing transferring and multiplying messages – Artificial manipulation of industrial refrigeration – magnifying transformers and watt-meters – Rotating field motors.

Beneath Tesla's flourish of a signature is a picture of his hands holding an X-ray plate, below which is written: 'Burning Atmospheric nitrogen by high frequency discharge twelve million volts.'

The document is a desperate attempt by forty-eight year old Tesla to recover his self-respect and redeem his public reputation. It lists all ninety-three patents that had been issued in his name. The front page shows his dreams for the future; the back page pictures the reality of his achievement at Niagara Falls.

No longer was Tesla able to afford the luxury of saying, 'I cannot be bothered with this trivia, I have great schemes to complete.' He was now forced to say to the electricity industry:

I wish to announce that in connection with the commercial introduction of my inventions I shall render professional services in the general capacity of consulting electrician and engineer . . . I shall undertake

the experimental investigation and perfection of ideas,
methods and appliances, the devising of useful expedi-
ents and in particular, the design and construction of
machinery for the attainment of desired results. Any
task submitted to and accepted by me will be carried
out thoroughly and conscientiously.

As well as circulating this statement to possible clients,
Tesla also published it in the February 1904 edition of
Electrical World and Engineer. No wonder he wrote to
Morgan that he was undergoing the sufferings of Job
in his attempts to save his vision of the future, but his
sufferings were not yet over. Tainted by Morgan's public
failure to support him, little interest was generated by
his advert.

Once again, he approached Morgan, this time suggesting that he be released from all obligations and that
the assignments of patents be returned to him so that he
might exploit them in order to repay Morgan's $150,000.
Morgan replied that a deal was a deal; he had stuck to
his part, now it was up to Tesla to keep the bargain.
The truth of his situation was finally dawning on Tesla
and, on 19 December 1904, he wrote an angry letter to
Morgan:

You say that you have fulfilled your contract with
me. You have not. When we entered our contract I
furnished patent-rights, my ability as an engineer and
electrician and my good will. You were to furnish
money, your business ability and your good will. I

WIRELESS POWER

> *have assigned patent rights which are worth ten times*
> *your investment . . . You discredited me.*

In February 1905, Tesla made a final proposal to JP Morgan, suggesting that Morgan exchange his patent interests for one third of the stock in a new company which Tesla planned to form. Morgan replied he would be happy to invest in a new Tesla company, provided the investment was secured but he did not wish to relinquish his control of the 'wireless' patents. Tesla's dream of free wireless power was dead, killed by JP Morgan's cunning. Now it could not threaten the AC power system that underpinned the thriving electricity industry.

Tesla never spoke openly about the treatment he received from Morgan, so the extent of his disillusionment can only be gleaned from his letters. In public, he remained a gentleman and, even though Morgan had effectively ruined him, could still find it in himself to say something good about the man. When, at the age of seventy, he remembered Morgan in an interview, he said, with the benefit of many years' hindsight:

> *Various rumours have reached me, that Mr. J. Pierpont*
> *Morgan did not interest himself with me in a business*
> *way, but in the same large spirit in which he has*
> *assisted many other pioneers. He carried out his*
> *generous promise to the letter and it would have*
> *been most unreasonable to expect from him anything*
> *more. He had the highest regard for my attainments*
> *and gave me every evidence of his complete faith*

*in my ability to ultimately achieve what I had set
out to do.*

*I am unwilling to accord to some small-minded and
jealous individuals the satisfaction of having thwarted
my efforts. These men are to me nothing more than
microbes of a nasty disease. My project was retarded
by laws of nature. The world was not prepared for it.
It was too far ahead of time, but the same laws will
prevail in the end and make it a triumphal success.*

But reading between the lines, his bitterness had not
totally healed even then.

The World System was a discredited failure and, at
the age of fifty, Tesla was again a penniless immigrant
with only his wits to support him.

chapter 12

The Turbine, the Nobel Prize
and the Edison Medal

I misunderstood Tesla. I think we all misunderstood Tesla. We thought he was a dreamer and a visionary. He did dream and his dreams came true, he did have visions but they were of a real future, not an imaginary one.

Charles A Terry, speaking at the award of
the Edison Medal to Tesla in 1917.

Fifty is not a good age to start again in a completely new field, but Tesla had no choice. Once again he was broke; JP Morgan controlled all his 'wireless' patents, and Westinghouse controlled all interest in wired AC power. All that was left for Tesla was the writing of 'wild' predictions about the future which he could not deliver. His writer's persona was steadily wrecking his image as an engineer, and, now that Morgan had pulled the plug on his World System, he had lost all financial credibility.

On his fiftieth birthday, Tesla sat alone in the Delmonico restaurant pondering his future. As a younger man he had made it his habit not to celebrate his birthday, claiming that because he was born on the stroke of midnight between two days he had no birthday. But on his fiftieth birthday, he had nothing to celebrate.

Fifty is often a time for looking back, and I wonder if Tesla was tempted to think back to his childhood as he waited for his meal to arrive? Perhaps he remembered the time when he was a champion crow-catcher before a really nasty experience made him give up hunting birds. His method for this had been simple. Going into the forest, he would hide in the bushes, imitate the call of a crow and keep calling until a crow came close. Next, having thrown something to distract the bird, he would jump on it before it could fly out of undergrowth. He caught many crows in this way and began to believe his method was infallible. One day he succeeded in catching two birds and, having picked them up by their feet, was carrying them home from the forest. The captured crows, however, made such a frightful racket that a large flock of crows assembled to mob him. At first, Tesla was amused, but when he received a blow on the head that knocked him over he realized it wasn't funny. The flock attacked him so viciously that he had to release the two captured birds and hide in a cave until the rescuers dispersed.

On his fiftieth birthday, he needed another cave to hide in – this time from a flock of competitors who had seen a chance to destroy him.

He had brought many of his troubles upon himself: he had little interest in money; his childlike belief in authority figures was hardly dented even when they were acting directly against his interests; his pride never allowed him to admit that a task was impossible; and lack of money had forced him to take up writing about visions of the future that could be sensationalized and hyped up, and used by other writers to support wild schemes or to ridicule his views. The angle didn't matter to the publishers, but the lurid coverage was beginning to isolate Tesla from the mainstream of engineering.

The publication of 'The Problem of Increasing Human Energy' which had founded his notoriety and brought about his disastrous business venture with JP Morgan, had been used to confirm his instability and, from then on, all his statements were regarded with suspicion. Whatever he talked about, the result was the same: 'Oh, it's just old Tesla sounding off again!'

Despite this, as he got older, he became quite a prolific writer. Why? He had more spare time, engineering jobs were hard to find, and he needed money to live. But there is an another less charitable reason. His advert, selling himself as a consultant engineer, had been unsuccessful and he was left feeling that the world neither understood nor appreciated him. He had always felt unappreciated by his parents while they were alive, now he felt the same was true of the world. His advert had been an attempt to justify himself as a respected inventor and a trustworthy engineer, and its failure to produce any new patrons or bring back previous supporters must have hurt him deeply. Likening his own suffering to that of Job is something only a deeply hurt man would do. Job had felt abandoned by God; Nikola felt abandoned by world. He continued to write to prove himself to that uncaring world.

Never having married, he had never enjoyed the luxury of being loved and accepted for himself, rather than for what he had achieved. What self-worth he had, had come from work and working relationships. Even so, he had few male friends and none of these was really close. Samuel Clements (Mark Twain) was probably the nearest he had to a confidant, but even that friendship had stemmed from his own hero worship of the writer, and I suspect that it meant far more to Tesla than to Clements. Tesla probably saw Clements as yet another father figure; and the fact that Clements had become very outspoken before his death in 1910, may have encouraged him to behave in the same way. The difference, however, was that Clements could afford to be outspoken because his work as a writer was not dependent on capital investment or backers. Tesla, on the other hand, could only carry on his research if financiers trusted him; and business backers do not see outspokenness as a virtue.

Tesla seemed to be attracted to writers because after Clement's death he became friendly with Robert Johnson, the editor who published that first 'outrageous' article 'The Problem of Increasing Human Energy'.

In 1905, Tesla made a last attempt to justify his World System in print. He wrote an extremely verbose complicated article entitled 'The Transmission of Electrical Energy Without Wires As a Means of Furthering Peace'. (*Electrical World and Engineer* must have paid him by the word as it is so prolonged and rambling, and he was so short of money!) The article, as this opening paragraph shows, is difficult to understand:

> *UNIVERSAL PEACE, assuming it to be in the fullest sense realisable, might not require aeons for its accomplishment, however probable this may appear, judging from the imperceptibly slow growth of all great reformatory ideas of the past. Man, as a mass in movement, is inseparable from sluggishness and persistence in his life manifestations, but it does not follow from this that any passing phase, or any permanent state of his existence, must necessarily be attained through a stataclitic process of development.*

Tesla then continued with the nature of peace, a discussion relating to communications, a brief flirtation with the idea of imposing a single language on all people, a consideration of how distance can be eliminated by improving transport, the under-exploited properties of electricity, and how important it is to have a cheap means of transmitting power. The use of phrases such as, 'I performed the great experiment on that unforgettable day that the dark God of Thunder mercifully showed me his vast, awe-sounding laboratory' may have won him fans among followers of the occult, but they did nothing to salvage his reputation as a rational engineer.

Reading this article with the vision of hindsight, it seems to be the outpourings of a man near to the end of his tether, making a last drastic attempt at self-justification as he sees his dream slipping away. Why else would he end the piece on:

> It is not a dream, it is a simple feat of scientific electrical engineering, only expensive blind, faint-hearted, doubting world! Humanity is not yet sufficiently advanced to be willingly led by the discoverer's keen searching sense. But who knows? Perhaps it is better in this present world of ours that a revolutionary idea or invention instead of being helped and patted, be hampered and ill-treated in its adolescence, by want of means, by selfish interest, pedantry, stupidity and ignorance; that it be attacked and stifled; that it pass through bitter trials and tribulations, through the heartless strife of commercial existence. So do we get our light. So all that was great in the past was ridiculed, condemned, combated, suppressed only to emerge all the more powerfully, all the more triumphantly from the struggle.

This article is a perfect example of Tesla's name being used by publishers for the sole purpose of selling magazines. The editor did Tesla no favours in accepting this almost unreadable exercise in self-justification, and the only motive can have been that Tesla's name on the cover sold more copies. In the event, the article simply added to Tesla's growing reputation of wildness and unreliability, and he certainly didn't emerge 'triumphant from the struggle'. Reading it must have confirmed Morgan's view that he was justified in side-tracking this brilliant but unreliable madman before he brought the electricity supply industry down around their ears.

The article also reopened old wounds with Edison. In an interview for the *New York World*, Edison commented that he did not believe that Tesla would now be able to talk around the world, but thought that Marconi would, sooner or later, perfect his system. The whole electrical world had turned on Tesla and disowned him. The crows were gathering again to attack him, but the cave he hid in this time had a stream of water running through it.

Tesla had been fascinated by the water ever since he had first experienced its power as a child. He had never forgotten that near-fatal accident when he was almost swept over the top of a high dam by the force of a river's flow, and had continued to experiment with it even while he was developing wireless power. Indeed, he had almost been killed again when a cast-iron cylinder he was using to pressurize water shattered and a piece of its shrapnel narrowly missed his head. He had survived, but the shattered cylinder had done considerable damage to the partly built Wardencliff laboratory.

Now, he remembered that, as a boy, he had built water-wheels as toys for his own amusement and that one of these had been strikingly successful. He had made a flat wooden wheel, cut as a slice across the diameter of a tree trunk and then shaped to a circle. Through its centre, he had put an axle that he suspended across a stream. The wheel had then turned with the flow of the stream and had worked for many months without stopping. From this simple boyhood experiment, Tesla now recalled that water flowing across a flat plate had made the plate turn, and that the buckets and fins used on a windmill or waterwheel were not necessary. Here was his chance to move into a new field: steam was used for most industrial processes and steam-engines used cylinders that moved backward and forwards, wasting energy; a bladeless turbine, on the other hand, could be very efficient. So Tesla decided to build one, persuaded a new acquaintance, a wealthy sugar refiner, to lend him some money and set to work.

His first turbine was a stack of very thin discs about 6 inches (15 cm) in diameter turning inside a close-fitting casing that had spacers between the moving discs. As steam passed through the casing, it was slowed by friction with the discs. That friction made the discs turn and they would go faster and faster until the friction was minimized. His first turbine was successful, it went so fast that the discs of the rotor were stretched by the speed of rotation and the size of the casing had to be increased by a thirty-second of an inch to allow for the expansion. At a speed of rotation of 35,000 revs per minute, the rotor had to be very carefully balanced to avoid shaking itself to pieces. Tesla's boyhood idea really did work!

Within four years, Tesla had made a 12 inch (30 cm) diameter turbine that developed 100 horsepower. He persuaded the General Electric Company to allow him to test his new turbine in a New York power station, the Waterside station that had first been built as an Edison DC power station but now generated AC using Tesla-designed generators. The publicity value of Tesla testing a radical new steam turbine on their premises, combined with the possibility that he might make a useful new tool for reducing the cost of electricity generation, persuaded GE to allow him his tests.

The venture did not run smoothly. Tesla, very much a night hawk, had taken to getting up later and later in the day and working long into the night. Indeed, he seemed to be turning night into day. While he was simply working in his own laboratory on Broadway this was not a problem, but when he wanted to use GE's plant and power station, it became one. The biggest load on the New York generating station was in the late afternoon and evening up to about midnight, whereas the load was much lighter during the day and a much easier time, therefore, to accommodate Tesla's experiments. But with the same lack of skill in personal relationships that he had shown throughout his life, Tesla insisted on arriving to work on his turbine at 5 p.m., and staying until midnight. No requests by the station staff to work during the day had any effect on him and, by the time the experiments were finished, everybody was glad to see the back of him.

He made two 18 inch (45 cm) turbines and tested them in a strange tug-of-war. The two 200-horsepower turbines were connected to a torque measuring shaft and set to rotate in opposite directions. The shaft stayed still, but the gauge connected to it showed how much energy the two opposing shafts were making. Fortunately for Nikola and his spectators the shaft proved strong enough, for if it had snapped the results would have been spectacularly lethal.

Some of the power-station workers who saw the experiment did not understand that the two turbines were supposed to stay still for the test and told their colleagues that they were a complete failure. These stories did nothing to restore Tesla's reputation, and as, by now, he had run out of money again, he needed an industrial partner. Allis Chambers of Milwaukee already made conventional turbines, so Tesla decided to approach this company with his ideas.

But instead of showing some sensitivity and approaching the engineering staff at Allis Chambers, he used his past reputation to arrange an audience with the company chairman, and convinced the man that his new turbine was the answer to his company's prayers. This going over the heads of the engineering staff won him no friends, and did nothing to ensure a favourable reception for his new ideas. As a result, when two test steam turbines were built and tried out, his colleagues' test reports were scathing, highlighting the distortion of the discs during normal operation and Tesla's odd way of working without any engineering drawings. The report to the Board said:

When the units were dismantled the discs had distorted

> *to a great extent and the opinion was that these discs*
> *would ultimately have failed if the units had been*
> *operated for any length of time. The gas turbine*
> *was never constructed for the reason that the company*
> *was unable to obtain sufficient information from Mr*
> *Tesla indicating even an approximate design that he*
> *had in mind.*

In reaction, Tesla again demonstrated his supremely poor interpersonal skills by walking off the job, with the comment, 'They would not build the turbines I wished.' And this was the end of his attempts to create a new career for himself in steam turbines. His design, however, was basically sound and his ideas are now used in modern gas turbines.

Tesla's immature urge to please authority figures also showed when the *New York Times* printed a false rumour that Edison and Tesla were to share the Nobel Prize for Physics. They were both interviewed. Edison simply said he knew nothing about it and did not wish to comment. Tesla gave a full interview, speculating at length that the world was finally recognizing the importance of his discovery of how to transmit electrical energy without wires. Warming to the topic, he added:

> *We can illuminate the sky and deprive the ocean of*
> *its terrors! We can draw unlimited quantities of water*
> *from the ocean for irrigation! We can fertilise the soil*
> *and draw energy from the sun! In a thousand years,*
> *there will be many recipients of Nobel prize. But I have*

not less than four dozen of my creations identified with my name in the technical literature!

He must have felt really foolish when the prize went to WH & WL Bragg. Edison, well aware that Nobel laureates are notified personally before the Press is informed, was wise not to comment.

When Tesla did finally win recognition, by being awarded the Edison Medal by the American Institute of Electrical Engineers in 1917, it was an ironic double-edged victory. The Edison Medal is a regular award, set up by a group of Edison's admirers to mark achievement in electrical science. The original terms of reference said that it should be given for the best graduating thesis by a student of electrical engineering in the United States and Canada, but for four years nobody applied for it. The terms of the medal were then re-drafted so that it could be awarded to any living resident of the United States, its dependencies, or Canada for outstanding achievement in electrical engineering. The first six medals were given to Elihu Thomson, Frank J. Sprague, George Westinghouse, William Stanley, Charles F. Brush and Alexander Graham Bell. In 1917, the awarding committee decided to give the seventh medal to Nikola Tesla.

The award meeting was held at the Engineering Societies Building in New York City on Friday 18 May. The President of the Society, Mr W. W. Rice, Jr., opened the meeting at 8.30 p.m. Full minutes were taken, including the citation and Tesla's response.

Numerous legends have arisen about this award. Some writers say that Tesla did not attend the meeting because he would not be associated with Edison; others claim he left the meeting early to feed pigeons in the park. The minutes of the American Institute of Electrical Engineers, however, indicate that he was at the meeting throughout, and that he responded warmly and at great length to the honour bestowed on him.

On the night, according to the minutes, Dr Kennelly, the Chairman of the Edison Medal Committee, explained what the Edison Medal was and what it stood for, then President Rice called on Mr Charles A. Terry, to say something about the struggles and the early work of Mr Tesla. 'I think there is a peculiar significance in the fact that Mr Tesla is to receive the seventh medal — the seventh in most calculations is considered a most excellent number to have,' Charles Terry said. He then spoke of how men of vision and intelligence have discovered things of value for the benefit of their fellows and told of great examples from the past, naming Michelangelo, Galileo, Sir Christopher Wren, Livingstone, Newton, Franklin, Westinghouse and, of course, Edison.

Mr Terry, who was clearly a long-winded man went on to say, 'Although hope of reward may and properly should exist as an added impulse to such endeavours, the chiefly effective force compelling to the long hours of hard work and personal sacrifices of such men is the "I must" which speaks from within the soul, and with our truly great men the desire for reward is better satisfied by a consciousness of achieving their aims and by the just commendation of their fellows than by material gain, except insofar as the latter may aid in the further advancement of their tasks.'

He then made the observation that good engineers are not envious of doers of great deeds, but are grateful and pleased to be able to mark such achievement – just as all the members present were pleased to honour Mr Tesla, the very same Mr Tesla who had stood before them twenty-nine years earlier to deliver his first paper on Alternate Current Motors and Transformers.

Summarizing Tesla's achievements, he said that 'It remained to the genius of Tesla to capture the unruly, unrestrained, and hitherto opposing elements in the field of nature and art and to harness them to draw the machines of man,' and praised Tesla's imagination, hard work, staying power and range of his successes. Turning to Tesla, he concluded his speech with: 'It is not possible in this brief survey even to touch upon many of the lines of Mr Tesla's activities, but we must content ourselves with this inadequate presentation of typical evidences of the fascinating genius of this man whom we delight to welcome as a citizen of our country – the country which he twenty-five years ago adopted as his own.'

Finally, extending his hand to commend Tesla to the members he added, 'Mr Tesla, we would indeed be woefully lacking were we not most cordially appreciative of your work, work which we know is good.'

Applause resounded throughout the hall and Tesla

gracefully, and with a modest smile, acknowledged the praise of the Institute members. The President then asked Mr B. A. Behrend to address the Hall.

> *Mr Chairman: Mr President of the American Institute of Electrical Engineers: Fellow Members: Ladies and Gentlemen: By an extraordinary coincidence, it is exactly twenty-nine years ago, to the very day and hour, that there stood before this Institute Mr Nikola Tesla.*

He then repeated the story of Tesla's career and waxed warm in his praise:

> *There is a time for all things. Suffice it to say that, were we to seize and to eliminate from our industrial world the results of Mr Tesla's work, the wheels of industry would cease to turn, our electric cars and trains would stop, our towns would be dark, our mills would be dead and idle. Yea, so far reaching is this work, that it has become the warp and woof of industry.*

He then moved into a lengthy technical description of Tesla's work, and could not resist emphasizing his own role in its stupendous development. Tesla, by now, was really enjoying himself. At last, at the age of sixty, after ten years of hard times and ridicule, he was getting the recognition he deserved. Moments such as these were certainly to be savoured.

THE TURBINE . . .

'We ask,' Behrend said, 'Mr Tesla to accept this medal. We do not do this for the mere sake of conferring a distinction, or of perpetuating a name; for so long as men occupy themselves with our industry, his work will be incorporated in the common thought of our art, and the name of Tesla runs no more risk of oblivion than does that of Faraday, or that of Edison.'

In this, however, Behrend was to be proved wrong. As we now know, Tesla's name has been largely forgotten outside electrical engineering.

'You have lived to see the work of your genius established,' Behrend continued. 'What shall a man desire more than this? There rings out to us a paraphrase of Pope's lines on Newton: "Nature and Nature's laws lay hid in night": God said, "Let Tesla be, and all was light."'

Behrend sat down to thunderous applause which, again, Tesla enjoyed to the full. It was just past 10 p.m. when the President rose to address him:

> It is easy, I think, for engineers and scientists to take for granted things that have been done in years past. When we sit under an apple tree and see the apples fall, it is an obvious phenomenon of nature. We can understand the laws of gravitation, but to Sir Isaac Newton, many years ago, this phenomenon, which to us to-day is so simple, helped him to an act of creative imagination of the most extraordinary kind.

He then went on to give yet another summary of Tesla's

work, this time concentrating on the building of the Niagara Falls Power station and comparing Tesla to Faraday. He finished with:

> *Mr Tesla, you hear tonight the many compliments which have been paid to you, but they are not bouquets merely cast for the adornment of the occasion — they have been given with the sincere appreciation of the electrical profession, and we give this medal to you in recognition of this, with full appreciation of what you have done for us, and with great hope that you may continue to contribute to our profession in the future.*

As he sat down to great applause, the audience was anticipating the close of the meeting because, traditionally, the response of the medal recipient was brief, but Tesla, enjoying his moment, rose to speak:

> *Mr President, Ladies and Gentlemen — I wish to thank you heartily for your kind sympathy and appreciation. I am not deceiving myself in the fact, of which you must be aware, that the speakers have greatly magnified my modest achievements. One should in such a situation be neither diffident nor self-assertive, and in that sense I will concede that some measure of credit may be due to me for the first steps in certain new directions; but the ideas I advanced have triumphed, the forces and elements have been conquered, and greatness achieved, through the co-operation of many*

able men, some of whom, I am glad to say, are present this evening, inventors, engineers, designers, manufacturers and financiers have done their share until, as Mr Behrend said, a gigantic revolution has been wrought in the transmission and transformation of energy. While we are elated over the results achieved we are pressing on, inspired with the hope and conviction that this is just a beginning, a forerunner of further and still greater accomplishments.

His audience waited for his few final words of thanks, but Tesla only just starting went on to describe in technical detail these proposed greater accomplishments, spoke of his childhood, of the great age achieved by his uncles, aunts and grandparents (unaware of the subtle ageing of his audience!). He described his working methods, how he liked to use his imagination, his thoughts on life after death and 'how many articles have been written in which I was declared to be an impractical unsuccessful man, and how many poor, struggling writers, have called me a visionary. Such is the folly and short-sightedness of the world!'

A lonely man, with a captive audience, and nobody to discreetly tell him to sit down, he just went on and on.

One of the conditions of the Edison Medal was that it had to be awarded to a living person, and Tesla, taking this to heart, now explained that he was well qualified to receive the medal because, for a man of his age, he was extremely alive. He then listed in great detail every childhood mishap and illness which had threatened his life; and doubtless by now some of his audience were regretting that he had survived!

He criticized Edison's way of working and explained how his own methods were far better and would be a model for future engineers. He told how he had been asked by students of psychology, physiology and other experts about his ability to see visions: 'You might think that I had hallucinations. That is impossible. They are produced only in diseased and anguished brains. My head was always clear as a bell, and I had no fear. Do you want me to tell of my recollections bearing on this?' he asked of the gentlemen on the platform, who, jerking to sudden attention, felt unable to do anything but nod encouragement. He then rambled remorselessly on, commenting on which of his two elderly aunts was the ugliest; of how a madman called Lucka had run about his childhood village frightening children; how as a child he was bitten on his bare stomach by a raging gander while wandering naked about the farmyard; how he still had a mark on his stomach from the incident. Mercifully, he did not offer to show them the scar!

For his audience, the discourse must have been taking on a surreal aspect as he moved into a description of Serbian funeral rites. And, by now, those present on the platform must have been praying that he would run out of anecdotes before midnight. But, undeterred Tesla was now describing how he was able to travel in his mind by means of mental visualization; explaining how he didn't have a birthday because he was born at the very stroke of midnight; how his heart had migrated

THE TURBINE . . .

from one side of his chest to the other, to the amazement of the medical profession; how he had survived a night locked in a haunted chapel.

Surely, his by now restless audience must have been thinking, he was winding up, but, no! he now began the full history of his education, including a mime to illustrate the size of his physics teacher's feet. Then, for the fourth time that evening, came the full story of his working career, this time in tremendous personal detail, such as describing how he taught radio pioneer John Stone, how standing waves worked: 'I said to Mr Stone: "Did you see my patent?" He replied: "Yes, I saw it, but I thought you were crazy."' The audience were doubtless on Mr Stone's side, but Tesla, as thick-skinned as ever, went on to explain how he convinced Stone of his genius.

At long last, a full hour after he had stood up to make his brief word of thanks, Tesla said, 'To conclude, gentlemen, we are coming to great results, but we must be prepared for a condition of paralysis for quite a while. We are facing a crisis such as the world has never seen before, and, until the situation clears, the best thing we can do is devise some scheme for overcoming the submarines, and that is what I am doing now.'

On that last patriotic reference to the war effort against German U-boats he sat down. And doubtless fuelled by intense relief that he had actually finished, the audience gave him a round of applause. But it was not over yet! Mr Alfred Cowles, determined not to be robbed of his moment of glory, presented Tesla with copies of photographs taken in Colorado during the lightning experiment, and Tesla, rising to thank him, started to speak again. The groan of the audience is not, of course, minuted but is easy to imagine.

'I have learned how to put up a plant that will develop a tension of 100,000,000 volts and handle it with perfect safety. This plant,' he said, waving the photographs at his now almost mutinous audience, 'was in Colorado. If anybody, who had not been dabbling in these experiments as long as myself, had done such work, he would surely have been killed.'

Remarkably he was not killed that night either, even when he droned on with tales of his Colorado experiments for another fifteen minutes. When he sat down again, the President rose hastily to his feet: 'If there is no further business, we will consider this meeting adjourned.'

The minutes, which were taken down in shorthand by the secretary Mr Hutchinson, ran to over 16,000 words. Tesla's reputation as a highly skilled but uncontrollable eccentric was now firmly consolidated.

THE TURBINE . . .

Top Secret Oblivion

The spread of civilisation may be likened to a fire;
First, a feeble spark, next a flickering flame, then a
mighty blaze, ever increasing in speed and power.

Nikola Tesla

Back in the 1960s, all young electrical engineers habitu-ally practised walking around with one hand placed in their pocket. This was not because they were slovenly or making an anti-establishment statement, but because of a warning they were always given in the first practise labs they attended: 'If you get an electric shock across your chest, it will kill you; get the same shock down one side of your body and it will just give you a jolt.' Thanks to this advice, electrical engineers who want to stay alive automatically put one hand in their pocket whenever they are near live electricity.

Nikola Tesla's Colorado Springs notes reveals that he was the first engineer to advise this safe working practise and, as a result, many electrical engineers owe him their lives. He also invented the automobile speedometer, the mechanical rev counter, radio broadcasting, AC power and the bladeless turbines.

How was such a versatile talented man, whose inventions make our modern civilization possible, forgotten? The names of his contemporaries, Edison, Marconi, Westinghouse and even JP Morgan, all became legends and live on, but Tesla is largely unknown to a public who still benefit from his works.

The scientific community has honoured him, and his name has been given to a unit for measuring magnetism. In one way, this is a fitting memorial, because he has been placed in the same hall of fame as Volta, Ampère, Gilbert, Henry, Hertz, Ohm and Faraday, great scientists who have all had electro-magnetic units named after them. But, although he has achieved this recognition by the informed, I can't help thinking that he would also have liked a more popular accolade.

After all, the quality of our modern life depends on a constant supply of electricity and it was his vision that made this possible. Yes, some engineers know his name, are taught it as a unit for measuring magnetic flux, but few know the story of the man who invented our twentieth century, and most hardly remember our debt to him.

I was a small boy, fascinated by electricity and desperately wanting my own wireless set, when I first heard about Nikola Tesla. I didn't want just any old wireless, I wanted an ex-navy AR88 radio receiver for my very own and spent many Saturday afternoons haunting the numerous second-hand radio shops of Manchester, searching for this coveted instrument.

I had friends who had their own wireless sets and was sometimes allowed to use my parents' large radiogram in the sitting-room, but that was not the same as having my own set. Tuning the radiogram to the very bottom of its tuning dial, right down below Radio Luxembourg, I could hear people who had their own wireless stations, talking to each other about the vast distances their radio short waves could go and bragging about the distant operators they could talk to. I wanted to know more – and curiosity drove my quest for an affordable short-wave radio of my own.

Wandering from shop to shop, carefully guarding the pocket that held my small savings, I sifted through pile after pile of junk and spent hours looking longingly at unaffordable new radios. Late in the day of one Saturday's unsuccessful search, I saw something that looked out of place on a dusty shelf – a gleaming polished lid of a wooden box which I just had to open and investigate. And this was how I came to spend all my carefully hoarded pocket-money on a Tesla Therapeutic Electrotherapy Machine.

As I carefully opened the lid, an accumulation of dust tickled my nostrils and made me sneeze. Looking inside, the machine, protected by a red velvet lining, seemed to be complete. There was a gleaming coil of enamelled wire, two copper cylinders connected to flexible wire leads, a brass switch and the empty space where a big battery went. The faded label on the underside of the lid praised the virtues of the high-frequency currents that this strange contraption had obviously once produced:

> *The currents furnished by this apparatus are an ideal tonic for the human nervous system. They promote heart action and digestion, induce healthful sleep, rid the skin of destructive exudations and cure colds and fever by the warmth they create. They vivify atrophied or paralysed parts of the body, allay all kinds of suffering and save annually thousands of lives.*

The wireless set, I decided, could wait a while. Here was a real piece of electrical magic! Riding home with it on the electric train, I dreamed of the experiments I would do and, in the event, that crude but effective electric-shock machine was the source of much childish satisfaction. Once I had a new battery, a large six-volt lantern battery that had to be specially ordered at the local radio shop, the coil buzzed and sparked in a spectacularly satisfactory manner, and, holding the copper cylinders, one in each hand, produced a strange tingling.

Enthusiastic about the benefit of Mr Tesla's high-frequency therapeutic currents, I persuaded some of my friends to form a circle. We all then linked hands with the two copper electrodes of the machine closing our loop. The electric tingling spread from hand to hand round the ring, evoking squeals of surprise as it passed through our twitching limbs.

This early success was the start of my lifelong interest in the hidden workings of all things electric. Electricity, which has only become important to society fairly recently, is still a comparatively new thing. My generation was probably the first to take it for granted. For instance, in the south-west of Ireland there is a small valley still known today as the Black Valley because, until a few years ago, it was the only place left in Ireland without a public electricity supply. Only 100 years ago, the simplicity of lighting our homes by pressing a switch would have seemed impossible magic – and without electricity the world would have remained a much harsher darker place.

Ask any well-informed person, 'Who invented electricity?' and they will probably answer 'Michael Faraday.' Visit any power station and do the children's quiz in the reception hall, and you will be convinced that this must be the case. But Faraday didn't invent our electrically-powered world of today. He simply carried out experiments that showed that electricity and magnetism always appeared together and wrote an important book *Researches in Electricity*. The man who gave electricity to the world was Nikola Tesla – the man who made the electric-shock machine that so intrigued me as a boy – the man whose story this book celebrates.

How then did he come to die alone and poor in an hotel room? Why was his body not found for two days, so that the date of his death is as uncertain as he considered the date of his birth? Why do most of the beneficiaries of his inventiveness not know his name?

Partly, his lack of fame was his own creation. Unlike Edison, Westinghouse, Marconi and Morgan whose companies preserve their names and keep their achievements in view, Tesla left no such monuments. The general public, if it remembers him at all, only remembers him as an outrageous contributor to newspaper columns. Look at of some of the titles of his later writings:

> *Tesla's Tidal Wave to Make War Impossible. Sleep From Electricity. How to Signal Mars. Mr Tesla on The Future. Nikola Tesla Plans to Keep 'Wireless Thumb' on Ships at Sea. Wonders of the Future. Famous Scientific Illusions. Nikola Tesla Tells How We May Fly Eight Miles High at 1,000 Miles an Hour. Can Radio Ignite Balloons? Signals to Mars Based on Hope of Life on Planet. Interplanetary Communication. Chewing Gum More Fatal Than Rum, Says Tesla. Breaking Up Tornadoes. Nikola Tesla Tells How He Would Defend Ethiopia Against Italian Invasion. Sending Messages to Planets Predicted by Dr Tesla on Birthday.*

In the twenty-six years between getting the Edison Medal and his death in 1943, the public's view of Tesla changed completely. He ceased to be regarded as a serious engineer and became a 'wild' old man who predicted miracles. The fact that most of these miracles eventually happened never seems to have counted in his favour. He never understood how to deal with people, either as individuals or in a crowd. His wildest statements were always based on theoretical reasoning. Sometimes he explained his thoughts, at other times he played the 'Great Man' and expected his readers to accept everything he said. He often moved on to a new idea without ever completing his work on an earlier one, and this gained him a reputation as a 'butterfly' mind. Occasionally, however, the world did remember and honour him.

On his seventy-fifth birthday, for example, he made the cover of *Time* magazine when he was traced to the Grosvoner Clinton Hotel where he was living on the goodwill of the manager after having been evicted from previous hotels for non-payment of his bills. Soon after this, he left the Grosvoner Clinton without paying his bill and had to forfeit his luggage. The Great Depression had clearly not helped his finances, and he only found a 'home' for the rest of his life when the Yugoslavian Government, taking pity on its most famous son living in poverty, awarded him a small pension of $7,200 per year. Even so, he still had to change hotels regularly because he encouraged pigeons into his room by feeding them on his desk.

He often claimed that he had taken a vow as a child to devote himself to work and never to waste time on marriage. But, as he got older, more garrulous and less respected, he must have regretted the lack of a close family to provide him with an audience.

TOP SECRET OBLIVION

His sisters all died before him, and the only member of his family he saw in his final years was a nephew, Sava Kosanovich, whom he did not seem to get on well with. Obviously lonely, he took to befriending young male science reporters and ringing them up to talk for hours at all times of the day and night.

Starting to celebrate the birthday he claimed not to have, he hosted dinners he could not afford, taking over popular New York restaurants to feed reporters and then making them 'pay' dearly for their dinner by forcing them listen to his long speeches about the future. He remained physically active until his eighty-first year when, struck by a New York taxi cab while crossing the street, his health started to decline.

On this birthday, instead of speaking at a dinner party, he issued a written statement. Although this was soon after the auto accident, his mind was obviously still capable of mounting an attack on Einstein's theory of relativity:

> *I have worked out a dynamic theory of gravity in all details and hope to give this to the world very soon. It explains the causes of this force and the motions of heavenly bodies under its influence so satisfactorily that it will put an end to idle speculations and false conceptions, as that of curved space. According to the relativists, space has a tendency to curvature owing to an inherent property or presence of celestial bodies.*
>
> *Granting a semblance of reality to this fantastic idea, it is still self-contradictory. Every action is*

accompanied by an equivalent reaction and the effects of the latter are directly opposite to those of the former. Supposing that the bodies act upon the surrounding space causing curvature of the same, it appears to my simple mind that the curved spaces must react on the bodies and, producing the opposite effects, straighten out the curves. Since action and reaction are coexistent, it follows that the supposed curvature of space is entirely impossible — But even if it existed it would not explain the motions of the bodies as observed. Only the existence of a field of force can account for them and its assumption dispenses with space curvature. All literature on this subject is futile and destined to oblivion.

It is a great pity that Tesla never published his dynamic theory of gravity. Modern thinking about gravity suggests that when a heavy object moves it emits gravitational waves that radiate at the speed of light. These gravity waves behave in similar ways to many other types of wave. Tesla's greatest inventions were all based on the study of waves. He always considered sound, light, heat, X-rays and radio waves to be related phenomena that could be studied using the same sort of maths. His differences with Einstein suggest that he had extended this thinking to gravity.

In the 1980s he was proved to be right. A study of energy loss in a double neutron star pulsar called PSR 1913 + 16 proved that gravity waves exist. Tesla's idea that gravity is a field effect is now taken more seriously than Einstein took it. But, unfortunately, Tesla never revealed what had led him to this conclusion, never explained his theory of gravitation to the world. The attack he made on Einstein's work was considered outrageous by the scientific establishment of the time, and only now do we have enough understanding of gravity to realize that he was right.

Tesla subsequently went on to make another outrageous claim, and the following statement helped to consign him to total obscurity after his death:

> I have devoted much of my time during the year to the perfecting of a new small and compact apparatus by which energy in considerable amounts can now be flashed through interstellar space to any distance without the slightest dispersion. I am expecting to put before the Institute of France an accurate description of the devices with data and calculations and claim the Pierre Guzman Prize of 100,000 francs for means of communication with other worlds, feeling perfectly sure that it will be awarded to me. The money, of course, is a trifling consideration, but for the great historical honour of being the first to achieve this miracle I would be almost willing to give my life.

He didn't get the prize and never explained the work. The French Government never heard from him as events overtook them both. Hitler was starting to expand his influence in Europe and France was invaded by 1940.

The device Tesla was talking about was either an early

laser or a plasma gun to produce high-energy particles in the upper atmosphere. His Colorado notes show that he was aware of both possibilities, and these devices would have been a logical consequence of his lightning experiments.

In 1940, just after his eighty-fourth birthday, he gave an interview to the *New York Times* that was published on 22 September:

> *Nikola Tesla, one of the truly great inventors, who celebrated his eighty-fourth birthday on July 10, tells the writer that he stands ready to divulge to the United States Government the secret of his 'teleforce', with which he said, airplane motors would be melted at a distance of 250 miles, so that an invisible Chinese Wall of Defence would be built around the country.*

The article passed without comment by fellow scientists. By now his reputation for seeking publicity far outweighed his ability to be believed, and with Hitler's advances in Europe causing concern there were other things to worry about.

By 1941, the US had entered World War II and Tesla must have been concerned when his native land also fell to German invaders about this time. What was he to do about his 'Death Ray', as the popular paper had dubbed his 'teleforce' weapon? He wanted to give it to the US Government to help support both his adopted country and his homeland.

On 5 January 1943, Tesla rang the US War Department and spoke to a Colonel Erskine, offering him the secrets of his 'teleforce' weapon. Erskine, not realizing who Tesla was, assumed he was crazy, promised to ring him back and forgot about him. This was Tesla's last message to anybody. Quite ill by this time, his weak heart causing regular dizzy attacks, he was living in the Hotel New Yorker. On the evening of 5 January, he gave orders that he was not to be disturbed and went to bed. He often told staff to leave him undisturbed for two or three days at a time, but this was to be the last time he would be seen alive.

The story now unfolds like a bad thriller. Tesla died of heart failure some time between the evening of Tuesday, 5 January and the morning of Friday, 8 January. He was found by a maid on the Friday morning. His only known relative, his nephew Sava Kosanovich, a refugee from Yugoslavia who had fled to the US to escape the German invasion, was, like many other refugees, under observation by the FBI as a possible spy.

On the night of 8 January, Sava Kosanovich and two other men, George Clark, and Kenneth Sweezey (a young science reporter) went to Tesla's hotel room with a locksmith to open his safe. Kosanovich told the other two men he was looking for Tesla's will. Three assistant managers of the New Yorker Hotel and a representative of the Yugoslavian Consulate were present as witnesses. Sweezey took a book from the safe, and the safe was then re-closed with a new combination that was given to Sava Kosanovich. If Kosanovich found a will he never produced it because Tesla is recorded as dying intestate. (Kosanovich did, however, eventually collect together all Tesla's remaining writings and equipment, which is now housed in the Tesla Museum in Belgrade.)

On the same evening, Colonel Erskine called the FBI to tell them that Tesla had died and that his nephew, Kosanovich, had seized papers which might be used against the US Government. The FBI made an immediate inquiry in New York, confirmed that Kosanovich and others had entered Tesla's room with the aid of a locksmith and contacted the Alien Property Custodian to retrieve the items seized on behalf of the Government.

Mr Fitzgerald of Alien Property Control then went to the hotel and took away all Tesla's remaining property, which consisted of about two truckloads. The articles were then sealed and transferred to the Manhattan Storage and Warehouse Co. NY, where other Tesla effects, a further thirty sealed barrels and bundles, had been stored since 1934. The Alien Property Custodian then seized all Tesla's effects on Saturday morning, and called in naval authorities to make microfilm copies of all his papers.

The FBI also discovered that Tesla had stored an invention in a safe-deposit box at the Grosvoner Clinton Hotel in 1932, but when the agents tried to claim it, the hotel refused to release the contents of the box unless Tesla's unpaid bill was paid. The hotel did agree, however, to notify the FBI if anybody else tried to get at it.

FBI records state that Sava Kosanovich was trying to gain possession of Tesla's effects, and that it was concerned that Kosanovich might make this information available to the enemy. The FBI consulted the scientific advisor to Vice-President Wallace, and was told to lose no time in doing whatever was necessary to preserve Tesla's effects. It was were also told that Tesla had completed and perfected his experiments in connection with the wireless transmission of power and had developed a new torpedo. The plans and a working model that cost $10,000 to build, were in the safety deposit box of the Grosvoner Clinton hotel. The model was connected with Tesla's Death Ray or the wireless transmission of electrical current.

The Bureau, ordered to keep the Vice-President informed of what actions it took, decided to approach the State's Attorney concerning the possibility of arresting Kosanovich on a burglary charge and therefore getting back the papers he had taken from the safe. At that point, the Alien Property Custodian took over responsibility for the securing of Tesla's property and the FBI record ends.

A memo was sent out from J Edgar Hoover instructing that 'all matters connected with the late Nikola Tesla are to be handled in a most secret fashion in order to avoid any publicity in respect to Tesla's Inventions and that every precaution be taken to preserve the secrecy of those inventions.'

So Tesla's life's work was declared TOP SECRET and discussion of it forbidden.

Ironically, Tesla's Death Ray was real, and it is only in the last few years that science has caught up with him. On 18 October 1993, the US Department of Defence announced it was starting to build an experimental ionospheric research facility in Gakona, Alaska. This facility, known as HAARP (High Frequency Active Auroral Research Program) was built by the Raytheon Corporation and involves the Universities of Alaska, Massachusetts, Stanford, Penn State, Tulsa, Clemson, Maryland, Cornell, UCLA and MIT in its program of experiments to study the resonant properties of the earth and its atmosphere. The link with Tesla's work is clear. HAARP is studying exactly the same phenomena which Tesla first considered nearly 100 years ago in Colorado.

HAARP is based on the ideas of Bernard Eastlund, who holds three US patents (4,686,605 – 4,712,158 – 5,038,664), all of which have been issued as improvements on the patents first issued to Tesla after his Colorado tests. The titles of the patents, which have to be shown to be practical before a patent is issued, are: method and apparatus for altering a region in the earth's atmosphere, ionosphere and/or magnetosphere; method and apparatus for creating an artificial electron cyclotron heating region of plasma; and method for producing a shell of relativistic particles at an altitude above the earth's surface.

This last patent, which describes an anti-missile shield which could destroy the electronics of hostile missiles or satellites, is the realization of Tesla's Death Ray. It works by creating a plasma packet of high-energy particles – Tesla's Colorado ball lightning on a large scale.

So, on two counts, Tesla had the last laugh: his 'teleforce' has finally been built and he won a patent battle with Marconi when, after he had been dead for six months, the US High Court confirmed it was Nikola Tesla who really invented radio! This was, of course, rather a hollow victory given that both patents had expired, both men were dead and nobody could talk about it because there was a Top Secret order forbidding discussion of Tesla's work.

The end result of this sad chain of events is that one of mankind's greatest benefactors is almost forgotten. Tesla died as he lived, alone, lonely and Top Secret, consigning himself to years of obscurity because of his last alarming offer to the US Government.

He was a scientist of dazzling brilliance, a prophet who really did see into the future but was unrecognized in his own time. He was such an individualist, so self-centred that he never formed a close relationship with anybody, man or woman. Yet he was enormously cultured, spoke many languages and was very well read. In his later years, he partly made his living by translating literature into Eastern European languages. But he never formed any lasting businesses or links with institutions which would have preserved a record of his achievements. The Tesla Museum in Belgrade was only established long after his death.

Any chance of celebrating his life's achievements was lost by the panic that his death caused in wartime USA

– his life's work filed as Top Secret by the FBI, US Navy and Vice-President Wallace, and it is only now, almost 60 years later, that we can remember him openly. He must have been difficult man to work with, his workaholic attitude and his failure to suffer fools gladly would have meant that lesser engineers suffered from his tongue. But what a splendid companion he must have been at the dinner parties that he held in his prime.

I sit here surrounded by this man's legacy: my electric-powered computer at my side in my study, lit by fluorescent electric light, heated by water pumped by an AC induction motor, listening to music broadcast on my mains-powered radio. As my scanner and Internet modem sit on the desk, ready to send and receive pictures and messages round the world, I am using Tesla's legacy.

As the sun sets over the Pennine hills, I look out on an array of Tesla's monuments carrying electricity around the country. In the distance I can see the megavolt cables of the National Grid strung between their high pylons as they hiss and crackle in the damp evening air. Across the valley runs a twin-stranded 11,000 volt local distribution line strung between its T-shaped wooden poles, and I can just see the transformer which drops the voltage to a safe 240 volts for the short cable run into my house.

When you next see a line of electricity pylons carrying the power that grants you a civilized life, put one hand in your pocket and spare a moment to thank Nikola Tesla, the lonely, forgotten, long-winded, obsessive, brilliant man who gave this to you.

Tesla summed up his own life in these few words:

I continually experience an inexpressible satisfaction from the knowledge that my polyphase system is used throughout the world to lighten the burdens of mankind and increase comfort and happiness, and that my wireless system, in all its essential features, is employed to render a service to and bring pleasure to people in all parts of the world.

select bibliography

Colorado Springs Notes, 1899–1900, Nikola Tesla, The Nikola Tesla Museum, Belgrade, 1978.

Edison: A Biography, Matthew Josephson, Eytre and Spottiswoode, Connecticut, 1961.

Edison: The Man who made the Future, Ronald W Clark, Macdonald and Jane's, London, 1977.

Experiments with Alternate Currents, Nikola Tesla, McGraw Publishing, California, 1904. Reprinted Omni Publications, California, 1979.

George Westinghouse, Francis G Leupp, McGraw Publishing, New York, 1918.

A History of Electric Light and Power, Brian Bowers, Peter Peregrinus Ltd in association with the Science Museum London, London, 1982.

A History of the Marconi Company, WJ Baker, Methuen, London, 1970.

Lectures, Patents, Articles, Nikola Tesla, The Nikola Tesla Museum, Belgrade, 1956.

Lightning in His Hand: The Life Story of Nikola Tesla, Inez

Hunt and Wanetta W Draper, Omni Publications, California, 1977.

My Inventions: The Autobiography of Nikola Tesla, Edited by Ben Johnson, Hart Brothers, New York, 1985.

Pioneers of Wireless, E Hawks, Methuen, London, 1927.

Prodigal Genius: The Life of Nikola Tesla, John J O'Neill, Neville Spearman, London, 1968.

Tesla Said, Compiled by John T Ratzlaff, Tesla Book Company, New York, 1984.

Tribute to Nikola Tesla (contemporary reviews of his work), The Nikola Tesla Museum, Belgrade, 1961.

Turning Points in American Electrical History, Edited by James E Brittain, IEEE Press, New York, 1977.

Newspapers, Magazines and Other Publications

Century Magazine; *Colliers*; *Electrical Engineer*; *Engineering News*; *Literary Digest*; *Newsweek*; *Time*; *North American Review*; *Scientific American*; *World Today*; *The Times* (London); *New York Times*; *New York Sun*; *New York Herald Tribune*; *Sunday Times* (London).

Minutes of the annual meeting of the American Institute of Electrical Engineers, held at the Engineering Societies Building, New York City, Friday Evening, 18 May, 1917, IEEE Press.

index